PAINL[illegible] Science Projects

Faith Hickman Brynie, Ph.D.

illustrated by Hank Morehouse

All inquiries should be addressed to:
Barron's Educational Series, Inc.
250 Wireless Boulevard
Hauppauge, New York 11788
http://www.barronseduc.com

Library of Congress Catalog Card No.: 98-4169
International Standard Book No. 0-7641-0595-7

Library of Congress Cataloging-in-Publication Data

Brynie, Faith Hickman, 1946–
Painless science projects / Faith Hickman Brynie ; illustrated by Hank Morehouse.
p. cm.
Includes index.
Summary: Provides guidelines for preparing a science project, including choosing a topic, doing research, collecting and testing data, and presenting the final product.
ISBN 0-7641-0595-7
1. Science projects—Methodology—Juvenile literature. [1. Science projects—Methodology.] I. Morehouse, Hank, ill. II. Title.
Q182.3.B79 1998
507.8—dc21

98-4169
CIP
AC

PRINTED IN THE UNITED STATES OF AMERICA
9 8 7 6 5 4 3 2 1

Acknowledgments

I wish I had room to list the names of all those elementary, middle school, and high school students who—through their serious attempts to master the spirit and methods of science—taught me what this book needed to include. To each of my students who glowed with pride after "catching on" to graphing, variables, or probability, I extend my sincere gratitude.

In addition, I wish to thank those dedicated professionals and friends who contributed to this effort, especially Jerry Resnick, the principal of Clara Barton High School for Health Professions in Brooklyn, New York, and Melanie Krieger, who has guided students through often prize-winning Westinghouse research projects. I am grateful to Grace Freedson and Linda Turner at Barron's for offering me this opportunity and to Tammy Brynie and Peter Green for their continuing help, interest, and support. Most of all, I treasure my husband, Lloyd, and my daughter, Ann, who make all things possible in my wee corner of the world.

CONTENTS

INTRODUCTION

A science project is like the ocean:

- To some kids, it's a chance for a cool, refreshing swim.
- To others, the water looks deep, the undertow seems strong, and every wave threatens to drown them.

Whether you're a "science swimmer" or a "science sinker," this book is for you. It will show you how to carry out a project that's as much fun as a day at the beach. Soon, you'll find yourself splashing happily in that sea called science—no matter how wide, deep, and cold it may look to you now.

Chapter One tells you how to ask good science questions and shape them into promising plans. You'll examine different kinds of projects and discover some surefire ways to ruin your project every time.

Chapter Two provides some clues on how to find information about your topic. Chapter Three helps you shape your ideas into a logical design that's practical, safe, and affordable. In Chapter Four, you'll learn how to collect data and turn your numbers into tables, graphs, and useful statistics.

Chapter Five suggests ways to tell the world what you've found by using displays, written reports, and oral presentations. Chapter Six is packed with a dozen, dandy checklists that will help you keep your head above water as you swim through your project. The lists quickly review everything from materials and supplies to project safety.

For all you kids standing on the shore feeling either eager or afraid to dive in, this book is calling, "Come on in! The water's fine!"

CHAPTER ONE

Hit the Ground Running

A science project? What's that?

A science project is your chance to ask a question, but you won't look in books for the answer. Instead, you'll observe, build, and experiment—just as scientists do.

IDEA HUNT

All kinds of questions about the natural and physical world are fair game for science projects. Have you ever wondered

- how a pendulum clock keeps time?
- why farmers put fertilizer on crops?
- if Brand X cookie contains more chocolate chips than Brand Y?
- whether left-handed people are better artists than right-handed people?

- if you could build a robot to clean your room or program a computer to do your homework?

THINK FOR A MINUTE

. . . about everyday things you take for granted at home, at school, or in your community. What questions can you ask about them? How might you find answers—not by asking an expert or looking in an encyclopedia, but on your own?

Help! Where do I start?

You've already started if you've asked a question about the world in which you live. That's where science projects begin—with questions you want to answer. If you're stuck, though, don't worry. You can find ideas for science projects everywhere.

In a pocket notebook, jot down things you see, hear, and think about. Don't worry if your ideas are right for a science project. Just record them before you lose them.

Use your notebook to record things you think about or notice. You might write things such as

- That pond out by the highway. I wonder what lives there.
- People playing guitars in the subway. Why do the strings have different pitches?
- Leaves falling in the park. Why are some red, some yellow, some brown?
- Seeds in my grapes. Do green grapes have fewer seeds than purple grapes?
- Ketchup on my T-shirt. Will one detergent get the stain out better than another?
- On the news, they said eating broccoli is good for you. I wonder if my friends eat many vegetables.

Where else can I find ideas?

Visit your school's media center, your public library, or a college library (if one is nearby).

- Leaf through science magazines such as *Science World*, *Odyssey*, *Scientific American*, *Popular Science*, *Discover*, and *Psychology Today*. Read articles and reports that catch your interest.
- Skim science articles and columns in newspapers and popular magazines such as *Time* and *Newsweek*.
- Scan the card catalog (on computer in many libraries) for books about subjects that interest you.
- Dip into your library's collection of CD-ROMs and videos about science topics.

Caution—Major Mistake Territory!

Most libraries own books of preplanned science projects. Such books may help you get an idea of what you're aiming for, but don't just pick a project and copy it. You don't want to repeat someone else's work. You want to do your own!

Other good ways to find ideas:

- Think about labs you've done in science class. Can you use a technique you've learned to study something different? Can you take an investigation in a new direction?
- Use your favorite search engine to surf the Net for information on a topic that interests you.

- Draw from your experience, maybe even your troubles. For example, a student who was plagued with a persistent case of athlete's foot studied factors affecting the growth of fungi.
- Think about your interests. If you like art better than science, why not compare the fading of different kinds of paints? If you prefer music, maybe you'd like to compare the pulse rates of volunteers when they listen to Mozart and Meatloaf. If you prefer science fiction to science, do a survey to find out if women read more futuristic novels than men.

- Look at two things together that might interact in some interesting or unexpected way. For example, suppose your local car wash stands near a stream or river. You might test the water upstream and downstream from the car wash to see if the runoff changes the water quality.
- Check the reference and database offerings of some Internet service providers for science tidbits.

- While you are on-line, check out some of the science fair Web sites. For lots of links, start with:
  http://www.infotoday.com/MMSchools/NovMMS/cyberbee11.html
  Check also:
  http://www.halcyon.com/sciclub/kidproj1.html
- Think about careers. Does some particular kind of work appeal to you? Does science have anything to do with the job?
- Read on-line science magazines such as *Omni* and *Discover*.
- Watch TV shows and videos dealing with science topics.
- Talking is a great way to find ideas. Pick the brains of family, friends, and neighbors.
- Don't overlook failure as a source of inspiration. If your nail polish goes glumpy, find out why. If the bread fails to rise, look for a reason.

Winning friends and influencing people

The time will come when you'll want to seek expert advice about your project, but don't be in a hurry. Bugging busy professionals when you're not clear about what you're doing can't win you many friends in high places.

Curb your letter-writing and phone-calling urges, at least for now. Don't even consider asking scientists or organizations for everything they have on your topic.

Three (not very pleasant) outcomes are more likely than not:

- They ignore your request because you obviously haven't done your homework.
- You receive nothing, because the person or organization you contacted lacks the time and money needed to send out truckloads of printed material to every student who asks.
- You get an enormous pile of stuff that you aren't interested in, can't read, and won't use.

Know your subject inside out before contacting experts or professional societies. Seek help from scientists only after you've researched your subject thoroughly and know precisely what help you need.

Play by the rules

A basketball game isn't much fun without rules. Neither is a science project. Before you go much further, learn the rules. The requirements for your project depend on your situation.

- Some projects are class assignments. Teachers make the rules for these. Learn the rules, and follow them to the letter.
- Some projects are for science fairs or competitions. Many local fairs follow the rules of the International Science and Engineering Fair. Your teacher will give you a copy of these rules and help you stick to them.

The rules are often detailed. Check them for such essentials as

- the kinds of projects allowed;
- the form of written reports;
- the size and content of a display;
- permissions for research involving people or animals or the use of dangerous chemicals.

Know the rules and follow them exactly.
No excuses. No omissions. No exceptions.

BRAIN TICKLERS
Set # 1

Try to find at least 12 possible science questions in this scene.

You and your friends are playing basketball on an outdoor court behind an ivy-covered brick building. Snow dusts the boughs of pines and the bare branches of oaks, but you're sweating. The players' long shadows dance along the ground in the failing light of evening. Smoke rises from a distant chimney, and birds circle overhead. From a foul line chalked on the cracked concrete, you sink the perfect free throw!

(Answers are on page 39.)

NARROW THE FIELD

If you have a specific question in mind for your project, congratulations! You're off to a good start.

However, maybe you haven't traveled that far yet. Maybe all you know is that you're interested in a general subject such as

- pollution
- lungs
- fish
- motors

- electricity
- food

You must narrow such broad topics to something precise you can actually investigate. How? Follow four rules:

1. Get *specific.*
2. Get *more specific.*
3. Get *even more specific.*
4. If rules 1–3 fail, get *even more specific.*

Get specific

Imagine yourself using a telescope to bring a distant object into sharp focus. The more powerful the telescope, the closer you can get to your subject. That's the view you need for the best possible science project. For example:

Topic	Specific	More specific	Even more specific
pollution	pollution in the pond in the park	pollution from nitrate fertilizer runoff from the flower beds	How do the amounts of nitrate in the park's pond compare before, during, and after rains?
lungs	lungs of athletes and couch potatoes	lungs of eighth-grade girls who exercise and who don't	Does the lung capacity of eighth-grade girls vary according to the amount of exercise they get?
fish	goldfish in the home aquarium	How active are goldfish in the aquarium?	Does the temperature of the water affect how fast goldfish swim?
motors	motors in cars	gasoline versus diesel motors in cars	Do owners of diesel cars spend more on fuel per mile driven than owners of gasoline-powered cars?
electricity	electricity used for Christmas tree lights	electric Christmas tree lights wired in series and in parallel	Do lights wired in series use less electricity than lights wired in parallel?
food	food preferences, perhaps flavors of ice cream	ice cream flavor preferences of people of different ages	Do sixth, ninth, and twelfth graders prefer different flavors of ice cream?

Take a tip from journalists. A good way to get more specific is to ask yourself

- Who?
- What?
- When?
- Where?
- Why?
- How?

For example:

TOPIC: POLLUTION

Who? Who's polluting?
What? Exactly what kind of pollution?
When? Is the pollution worse at certain times?
Where? Is it worse in some places?
Why? Why is it a problem? What harm does it do?
How? How can it be decreased or stopped?

TOPIC: LUNGS

Who? Whose lungs? Boys or girls? Kids or grown-ups? Mice or hamsters?
What? What about those lungs? Their size? Their gas content?
When? Does lung capacity vary in people of different ages?
Where? Do lungs work harder at high altitudes?
Why? Why does lung capacity differ in males and females?
How? How can lung capacity be increased?

BRAIN TICKLERS
Set # 2

Write *Who?, What?, When?, Where?, Why?,* and *How?* questions about fish, motors, electricity, food, or a topic of your own. Then rewrite the following questions to make them more specific.

1. What makes clouds?
2. What's a good way to save energy?
3. What makes bridges strong?
4. How do cars affect air quality?
5. Can music calm people?
6. What makes milk spoil?

(You'll find some suggestions on page 40.)

Dump the impossible, but wait. . . .

As you get more specific, some of your ideas may seem impossible.

- "I want to do my project on monkeys," Melinda moaned, "but no way will they let me experiment on one."

- "I want to build a better bus," Benjamin blubbered, "but it won't fit in my display."
- "I want to put a space station in orbit around Saturn," Sarah shrieked, "but NASA won't foot the bill."

Let's face it. Some project ideas are too big, too time consuming, too expensive, or too ambitious to be practical. Before you dump that idea you love, though, look for another way to do the same thing. For example:

Instead of...	Try...
monkeys or elephants	beetles or mealworms
bus, plane, or train	models or toys
space station	model rocket
inheritance in humans	inheritance in Indian corn
human or animal hormones	plant hormones
full-size greenhouse, turbine, or windmill	miniature greenhouse, turbine, or windmill
gears	spools, pegs, Peg Board, rubber bands
meteor craters	rocks dropped into a pan of soil
cloning humans or sheep	cloning lettuce or African violets
animal breeding	plant breeding or grafting

Scaling down is another way of getting more specific. Ask yourself what you really want to know or do. Then scale your idea down to a manageable, affordable size.

BRAIN TICKLERS
Set # 3

What scaled-down alternatives can you think of for

1. trees?
2. washing machines (for testing detergents)?
3. mountains (for studying erosion)?
4. 747s?
5. the ocean?
6. lunar probes?
7. outer space?
8. the sun?

(Check page 43 for some suggestions.)

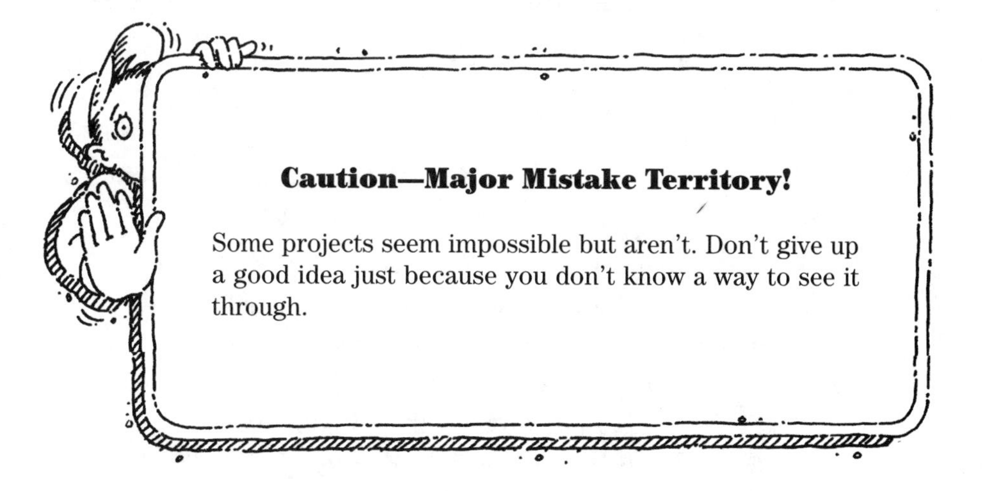

Caution—Major Mistake Territory!

Some projects seem impossible but aren't. Don't give up a good idea just because you don't know a way to see it through.

If your project idea seems impossible to carry out—even when scaled down—think again. Ask your teacher, your librarian, or an expert if he or she knows of a way to do what you want to do. Did you know, for example, that you can

- get tiny "bugs" out of the soil with a device called a Berlese funnel?
- order kits for testing air and water from scientific supply companies?
- build a simple tool to measure the angle of a slope?
- find the amount of organic (formerly living) matter in soil by baking and weighing the soil?
- use a microwave oven as a convenient source of radiation?
- (often) borrow equipment (such as a Geiger counter, cloud chamber, or spectrophotometer) from a high school, college, or commercial laboratory?
- capture, identify, and measure the amounts of invisible gases such as oxygen or carbon dioxide?

For more ideas, see the checklist on page 256, Chapter Six, "Supplies and Materials You Thought You Couldn't Get." "Simple Substitutes" on page 255 may help, too.

Definitely do dump

Some ideas should never get off the ground. Don't even think about

- performing surgery on animals;
- giving drugs or alcohol to humans or animals;
- asking people embarrassing questions;
- causing stress to people or animals;
- exploding anything or setting anything on fire;
- displaying firearms, drugs or drug paraphernalia, or preserved animals or body parts;

- experimenting with human or animal tissues such as blood and urine (unless you are working with a doctor or a scientist);
- using dangerous chemical or radioactive materials (unless you are working with an expert and have permission);
- removing plants or animals from their natural environments (except in certain cases—check with your teacher); or
- growing microorganisms that cause disease in plants, animals, or humans.

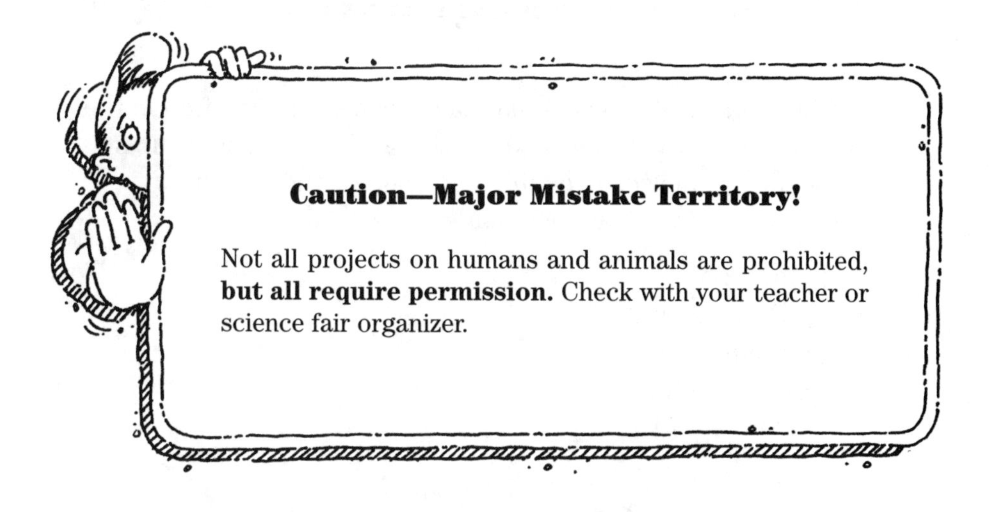

Science projects using possibly dangerous chemicals or equipment are permitted under certain circumstances and with help. Check with your teacher if you're thinking about using acids, lasers, large vacuum tubes, pressurized tanks, heaters, high-voltage wiring, pulleys, chains, isotopes, radiation, or other hazardous substances or devices.

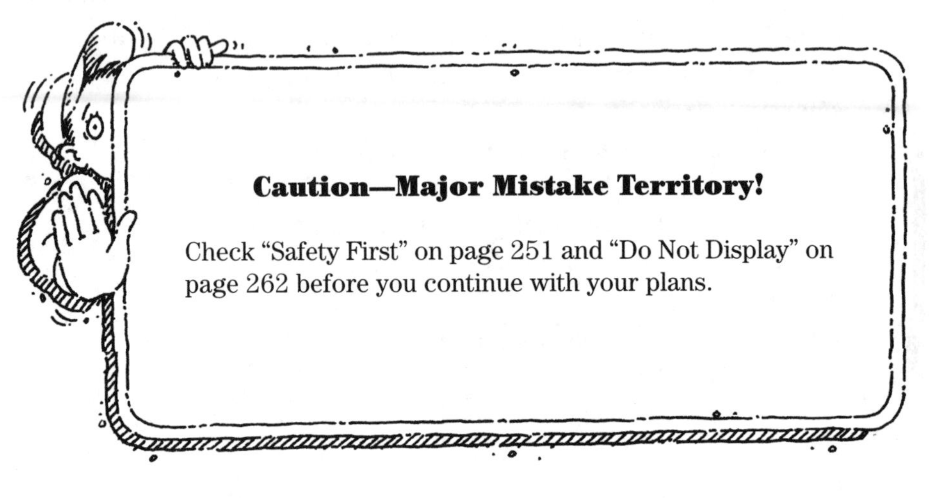

Caution—Major Mistake Territory!

Check "Safety First" on page 251 and "Do Not Display" on page 262 before you continue with your plans.

Steer clear of projects that have been done too often in the past. Models of volcanoes and of the solar system show up so frequently, some science fairs won't accept them. Think creatively. You want an idea that's

- original,
- exciting, and
- uniquely *yours*.

KINDS OF PROJECTS

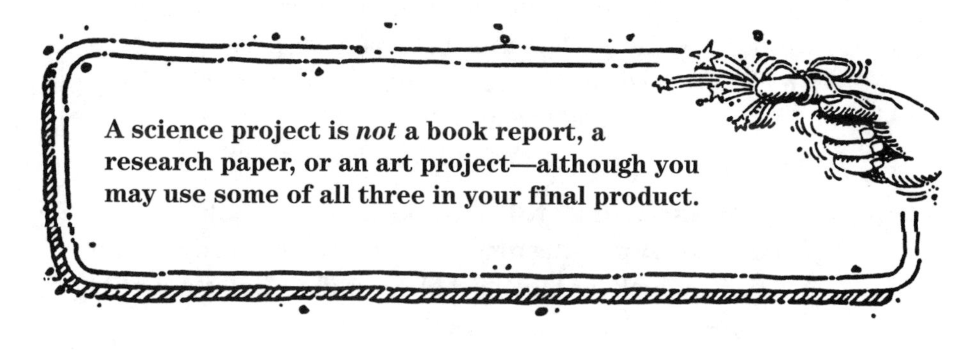

A science project is *not* a book report, a research paper, or an art project—although you may use some of all three in your final product.

Here's a new term to learn, use, and make your own:

Variable. Something that can change.

A variable changes because

1. you—the experimenter—change it on purpose (the cause), or
2. another variable causes it to change (the effect).

For example, turning on the stove under a pot of water (the cause) makes the water heat up (the effect). The on-off status of the stove is the variable you change on purpose. The temperature of the water is the variable that changes as a result.

The kind of project you do depends on the variables you study and how you treat them.

Most projects fall into one of three categories:

Non-experimental: You change no variables.
Almost experimental: You don't change any variables, but you look for the effect of a variable that is naturally there.
Experimental: You purposely change one variable and measure the result.

Caution—Major Mistake Territory!

The kind of project you choose depends on your question, but check the rules. Some teachers assign experimental projects exclusively. Some science fairs accept only engineering or computational projects.

Nonexperimental projects

In nonexperimental projects, you do not change any variables. However, you may measure one or more variables.

Demonstration or model

Models and demonstrations are good for

- teaching a scientific principle;

- showing how something looks or might look;
- explaining how something works.

You're considering a demonstration or model project if you've asked

- How can I show that Newton's second law of motion is really true?
- What should NASA put into an orbiting space station?
- How can I show that terracing reduces erosion?
- Any question that calls for a show-and-tell answer.

Observation

You're heading toward an observational project if you plan to

- record the behaviors of animals in the wild;
- collect and identify microorganisms from a pond;
- count the food choices of people in a cafeteria line;
- catalog or describe anything in its natural state.

Observation seems like the simplest kind of project, but it's probably the hardest to do well. Why? Observation requires a lot more than just looking. *Observation* means counting, recording, and looking beyond the obvious.

Survey

You're on your way toward a survey project if you want to

- test seventh graders' knowledge of rock groups and report the results;
- ask people which brand of soft drink they prefer;
- take a poll to predict who'll win an election;
- find out people's knowledge or attitudes about any topic.

Surveys aren't as easy as they look. You'll need to ask careful, complete questions. You'll also need good math skills to handle the results.

For many science fairs, survey projects—or any project involving human subjects—must be approved by a committee or review board. The student must submit in advance a research plan that includes a copy of the questionnaire, survey, or test proposed for use. The student must also define the risk to subjects and create a consent form that all subjects sign before participating in the survey. Months may pass before you receive permission. Plan ahead!

Almost experimental projects

Your project is almost experimental if you don't change any variables but you look for the effects of variables that change naturally.

Most not-quite-an-experiment projects look for correlations. *Correlation* means that two things vary together. As one variable changes, so does the other.

You may be planning a correlation project if you've asked

- Are more students absent from school in warm months than in cold months?
- Do girls do better in math than boys?
- Do more insect larvae live in the deep parts of the pond or in the shallow parts?
- Any question where you want to see if two things relate to each other (and if so, how).

Caution—Major Mistake Territory!

Just because two things change together doesn't mean one causes the other to change. For example, tall people tend to be heavier than short people, but you can't lose weight by shrinking your height.

Experimental projects

In an experimental project, you change a variable and measure the result.

Screening

A lot of medical, agricultural, and consumer research involves screening. *Screening* means testing a large number of choices to find the one that's best for some purpose.

You may have a screening project in mind if you are thinking about

- finding the most absorbent brand of paper towel;
- measuring the vitamin C content of potatoes cooked in different ways;
- counting the days before different kinds of radish seeds sprout;
- any project that compares different things by some *outcome measure* (what happens).

Design-redesign

This is trial-and-error research. Many engineering and computational projects fall into this category.

You may be on your way toward a design-redesign project if you plan to

- program a computer to correct the speech distortions of divers breathing helium;
- test different wing designs on model airplanes;
- invent a new kind of solar oven;
- test different ways of doing something new or better.

If the real thing is too big, too expensive, or too complicated, use a model for a design-redesign project. This use differs, however, from nonexperimental models built only to show how something looks or works.

The trick with design-redesign projects is to change *one variable at a time* and see what happens. Keep careful records of every change (no matter how small) and the results of the change.

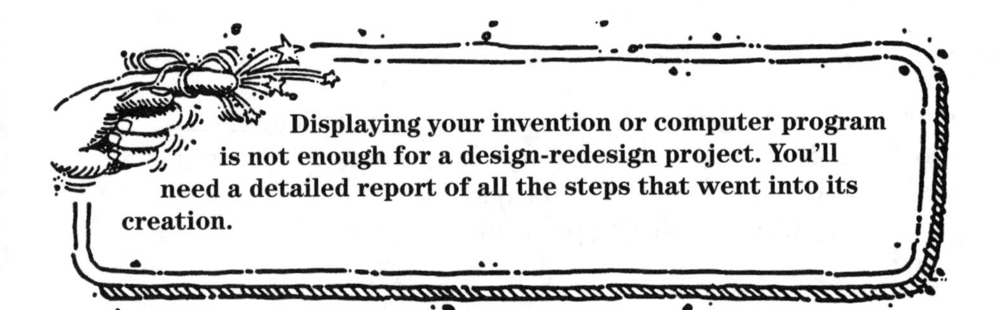

Displaying your invention or computer program is not enough for a design-redesign project. You'll need a detailed report of all the steps that went into its creation.

True experiment or fair test

To perform a fair test, the experimenter

- changes one or more variables in a planned way,
- keeps all other variables the same, and
- measures the outcome.

You've begun to design a true experiment if you've asked whether

- the height from which a baseball falls affects how high it bounces;
- bean seedlings grow taller under red light than green light;
- students remember history facts better if they study while listening to music;
- one thing can cause something else to happen when all other factors are kept the same.

BRAIN TICKLERS
Set # 4

Classify the projects below as one of these types.

A. Demonstration or model (nonexperimental)
B. Observation (nonexperimental)
C. Survey (nonexperimental)
D. Correlation (almost experimental)
E. Screening (experimental)
F. Design-redesign (experimental)
G. True experiment or fair test (experimental)

1. Do sweet peas grow tallest in potting soil, garden soil, or peat?
2. Can computer-assisted design be used to build a better mousetrap?
3. What species of birds feed at Coal Creek Pond?
4. Which potato chips contain the most fat?
5. Do hamsters prefer red food pellets to yellow ones?
6. Do left-handed students do better in foreign languages than right-handed students?
7. What kind of grocery bag rots fastest when buried?
8. How can a computer be programmed to analyze baseball statistics?
9. Which insulating material holds heat best?
10. Do adults and teens choose different situation comedies on television?
11. Do students who make good grades in math also make good grades in science?
12. How is $C_6H_{12}O_6$, the glucose molecule, constructed?

(Answers are on page 43.)

WHAT'S A HYPOTHESIS?

Did somebody say something about a *hypothesis*? No doubt someone did. You've already started thinking about yours, if you've

- thought of a way to measure your project's *result* or *outcome*;
- guessed how your test might turn out;
- predicted how one variable might affect another.

That's the definition of a hypothesis: the experimenter's educated guess about what will happen. (For mathematical reasons, scientists often use the null hypothesis: meaning, what they think *will* happen, *won't*.)

Outcome measures

Any project that's experimental or almost experimental uses some way of measuring the results. For example,

The height, dry weight, or number of leaves	measures	plant growth.
The number of colonies on an agar plate	measures	growth of bacteria.
The distance traveled in a specified period of time (10 seconds, 1 minute, and so on)	measures	speed or rate of travel.
A decibel reading (on a sound meter)	measures	how loud a sound is.
The f-stop reading on a camera's light meter	measures	how bright a light is.

Even nonexperimental projects have ways to measure outcome. Observational projects and surveys require *tallies* or *counts*. ~~||||~~ Ask yourself "How many?" and "How much?" for everything you do.

When planning your outcome measure, use the same rules you used to create your questions.

1. Get *specific*.
2. Get *more specific*.
3. Get *even more specific*.
4. If rules 1–3 fail, get *even more specific*.

Vague	Specific	More specific*
warmer	higher temperature	a temperature difference of x°C
bigger	heavier	a weight difference of x grams
bigger	taller	an average height of x centimeters
better	faster	moving at an average speed of x meters per second

better	more accurate	x = the average number of baskets made in every ten free throw tries
more	greater number	x = the average number of unpopped kernels per bag in popcorn Brands X and Y

*x stands for a number you won't know until you do the experiment.

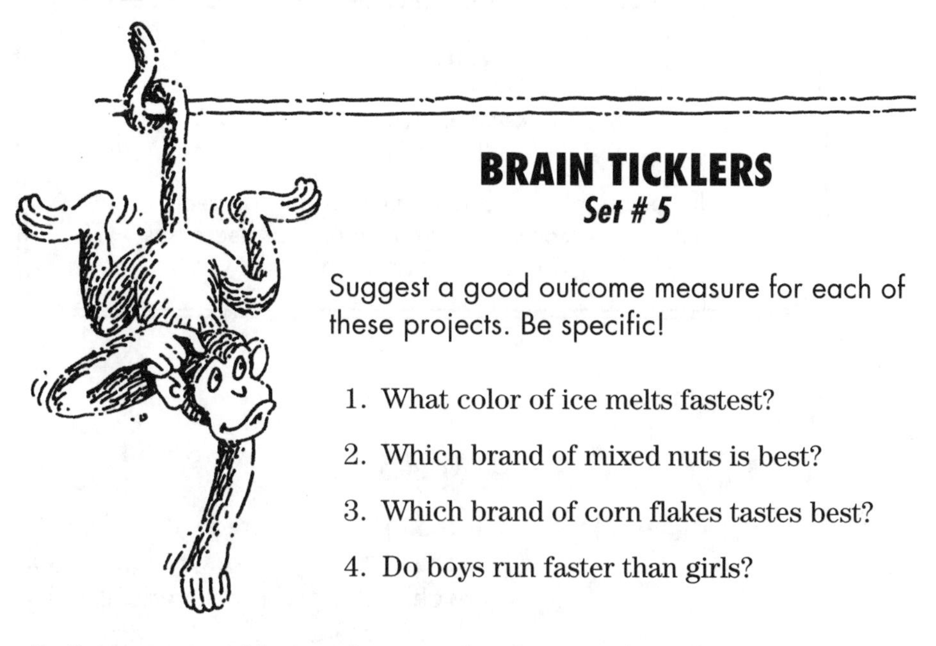

BRAIN TICKLERS
Set # 5

Suggest a good outcome measure for each of these projects. Be specific!

1. What color of ice melts fastest?
2. Which brand of mixed nuts is best?
3. Which brand of corn flakes tastes best?
4. Do boys run faster than girls?
5. Is one material better than another for parachutes?
6. Does my computer program work better than the shareware one I downloaded?
7. Do Canada geese live at Shiffer's Pond all winter?
8. Do apples float better than oranges? (Careful: This one is tricky!)
9. Can a plant be overwatered?
10. Who likes baseball better—teens or adults?

(Answers are on page 45.)

Guessing and predicting

A hypothesis is a guess or prediction of how the outcome measure will turn out. For experimental projects, it usually takes the form of an "if . . . , then . . ." statement. That means if one variable changes in a certain way, then the outcome measure will change as a result.

Hypothesis:

If **cause**, then **effect**.
If *A*, then *B*.

If the variable you change increases or decreases, then the outcome measure increases or decreases in a predictable way.

For example,

If . . .	Then . . .
fertilizer makes plants stronger,	bean plants watered with Wonder-Gro will have thicker stems after six weeks than those watered without Wonder-Gro.
sow bugs prefer cool places,	after 30 minutes, I'll count more bugs at the cool end of the temperature-gradient chamber than at the warm end.
short pendulums swing faster than long ones,	a pendulum with a 4-centimeter string will swing more times in 20 seconds than a pendulum with a 6-centimeter string.
hay fever makes kids miss school,	graphs of pollen counts and school absences will rise and fall together.

Not all hypotheses need to be "if . . . , then. . . ." For example, design-redesign hypotheses may simply state how changing one variable may affect the outcome measure. That's no excuse, though, for vagueness.

VAGUE:

A computer simulation will show that my car design is best.

SPECIFIC:

A computer simulation will show that a rear-wheel-drive car wobbles 10 percent less at 100 kilometers per hour than the same design with front-wheel drive.

Watch out for assumptions

Don't assume something is true without evidence. Don't accept *anything* without proof—not even something "everybody knows."

Spot the faulty assumptions in these projects:

- Which kind of music—classical or rock—affects plant growth the most? (Assumes music of any and all types affects plant growth.)
- Why are paper bags better for the environment than plastic? (Draws a conclusion without evidence.)
- Why aren't people more concerned about their health? (Assumes the results of a survey before it's ever conducted.)
- Why can't life exist in space? (It hasn't been found so far, but that doesn't mean it might not be found someday.)
- How come kids who eat breakfast get better grades? (You won't know if this is true among the kids you study until you collect data on breakfast-eating habits and grades.)

BRAIN TICKLERS
Set # 6

Write an "if . . . , then . . ." hypothesis for each of these projects.

1. How does the human eye respond to light and darkness?
   (outcome measure = pupil diameter)
2. Does potting soil hold more water than sand?
   (outcome measure = dry weight)
3. Does rock music make the heart beat faster?
   (outcome measure = pulse)
4. Do girls remember numbers better than boys?
   (outcome measure = a pencil-and-paper test)
5. Does country air have more dust than city air?
   (outcome measure = number of particles collected on a greased card)
6. Can seashells purify water?
   (outcome measure = amount of lead in water)
7. Do pea plants lose more water in light than in darkness?
   (outcome measure = dry weight)

(Answers are on page 46.)

Caution—Major Mistake Territory!

Some students forget that a hypothesis is a guess. They are so sure they know what will happen, they don't do the experiment carefully or they ignore their experimental results.

Keep an open mind. You can never prove that a hypothesis is right. You can only find that—so far—nothing has shown it to be wrong.

Progress check

By now you should have a pretty good idea of

- the question you want to answer;
- the kind of project you want to do; and
- how to measure and predict your outcome.

You may now be eager to learn more about your subject. That's in Chapter Two. However, before you go on. . . .

TEN SUREFIRE WAYS TO RUIN YOUR SCIENCE PROJECT

Rule #1. Pick a boring topic you couldn't care less about.

Rule #2. Write to experts asking for "everything you have on. . . ."

Rule #3. Cut all the corners you can. Get by doing as little work as possible.

Rule #4. Quit the first time something goes wrong.

Rule #5. Count on friends, parents, or teachers to bail you out at the last minute.

Rule #6. Make a pretty display. Who cares if it actually means anything?

Rule #7. Ignore the rules.

Rule #8. Don't bother to keep records or take pictures.

Rule #9. Do your experiment only once or just ask a few people a couple of questions. Isn't that enough?

Rule #10. Put it off. You can always throw something together the night before it's due.

BRAIN TICKLERS
Set # 7

Match each of these sad tales to a surefire way to ruin a science project.

1. Jennifer's teacher assigns a science project on October 10. It's due January 9. Jennifer figures she can do it over the winter vacation.
2. Roger wants to enter a computational science fair. He knows they want math and computer stuff, but he builds a cool house for crickets to live in instead.
3. Tamara drops a basketball from a height of 200 centimeters. It bounces to a height of 75 centimeters. She records those numbers. Her project is done.
4. Clint is interested in manatees, but what kind of project should he do? He writes to the Florida Manatee Society and asks for brochures.
5. Maura tries to borrow electrophoresis equipment from a nearby high school, but it doesn't have any. She scraps her idea and decides to do a book report about famous scientists.
6. Gerald paints his plant trays blue, builds a sturdy display of pine and cardboard, and buys stencils to make nice lettering. Then what should he do?
7. Sandra decides to study the burrowing habits of earthworms because her brother says she should and because she can't think of any other project.

8. Bill isn't worried about his science project. His dad's good at stuff like that.

9. Tabitha runs some toy cars down ramps and measures how far they travel, but she forgets to write down her results.

(Answers are on page 47.)

BRAIN TICKLERS—THE ANSWERS

Set # 1, page 10

You and your friends are playing basketball (1) on an outdoor court behind an ivy-covered (2) brick (3) building. Snow (4) dusts the boughs of pines and the bare branches of oaks (5), but you're sweating (6). The players' long shadows (7) dance along the ground in the failing light of evening (8). Smoke rises (9) from a distant chimney, and birds (10) circle overhead. From a chalked foul line on the cracked concrete (11), you sink the perfect free throw (12)!

1. Do some brands of basketball bounce higher than others? Do basketballs bounce higher than soccer balls?

2. Why does ivy cover the brick on the east side of the building but not the west?

3. Are all bricks the same? Do bricks differ in their strength or how much heat or water they hold?

4. Is all snow alike? Under what conditions does snow melt fastest?

5. Why do some trees lose their leaves in winter and others do not?

6. What causes humans to sweat even when the weather is cold?

7. How does shadow length relate to the length of the object that casts it?

8. How does shadow length change depending on the time of the day (or season of the year)?

9. Why is the smoke rising straight up instead of at an angle?

10. What birds live in this environment in summer? In winter? What birds migrate through this environment?

11. What causes concrete to crack?

12. What arc from the foul line guarantees a perfect free throw every time? (This is a good computer simulation question.)

Set # 2, page 14

These are only examples. You may have different questions that are just as good or better. The point is to **be specific!**

TOPIC: FISH

Who? What kind of fish (their name or species)? To whom are they important?
What? What factors affect the survival of the fish?
When? Are the fish more abundant in some years than others?
Where? Are these fish more abundant in some parts of the stream than others?
Why? Why are fishermen catching fewer trout in Cripple Creek now than they did five years ago. (Caution! This is a good question ONLY if you have lots of hard proof, not just an opinion, that the number of trout really has dwindled.)
How? How can the number of trout in Cripple Creek be restored (if you have proof that it has declined)?

TOPIC: MOTORS

Who? Who uses motors? For what purpose?
What? What kinds of motors are used in boats? How do they differ?
When? Do motors use less fuel when the weather is hot than when it's cold?
Where? Do boat motors get better fuel mileage in deep water than in shallow water? At high speeds or at low speeds?
Why? Why have car manufacturers changed carburetor designs over the last decade? (This is a good engineering question.)
How? How can motors be made more fuel efficient?

TOPIC: ELECTRICITY

Who? Who uses electricity? For what purposes?
What? What household appliance uses the most electricity? The least?
When? When is community energy use greater—night or day? In summer or winter?
Where? Where is electricity generated in this community?
Why? Why are electricity rates to commercial users lower than rates to households?
How? How is electricity generated in this community?

TOPIC: FOOD

Who? Who adopts a vegetarian diet? Why?
What? What food choices exist for vegetarians?
When? Does food spoil more quickly when it's left at room temperature than when it's refrigerated?
Where? Where do the ingredients in bread come from?
Why? Why do experts say most people should eat more vegetables?
How? How can seventh graders in this school improve their diets?

Vague	Specific	More specific
1. What makes clouds?	Is the humidity higher on cloudy or cloudless days?	Do temperature, humidity, and air pollution measures predict the percentage of cloud cover observed at noon every day for six weeks?
2. What's a good way to save energy?	How does lowering the setting on the hot water heater affect the family's energy bill?	How much energy do seven major home appliances use in an hour of operation?
3. What makes bridges strong?	Are suspension bridges stronger than trussed bridges?	In two model bridges of the same weight and height, will the suspension model or the trussed model support a heavier load before collapsing?
4. What about cars and air pollution?	Does my family's car emit nitrous oxide?	How much nitrous oxide does my family's car emit while idling for five minutes?
5. Can music calm people?	How does music affect blood pressure?	What is the average blood pressure reading of 20 people before, during, and after listening to a lullaby for five minutes?

6. What makes milk spoil?	How long before milk curdles?	After how many hours does milk curdle when left at room temperature as compared with milk stored at 4°C?

Set # 3, page 17

1. bean plants
2. baby food jars (shake them to simulate agitation)
3. soil in a see-through shoe box
4. paper airplanes
5. a jar of salt water or a saltwater aquarium
6. "robots" made with an Erector set
7. a vacuum pump and bell jar (borrowed from a physics lab)
8. a lightbulb

Set # 4, page 28

1. E. Screening, because you'll test three soils to see which grows the tallest plants.

   Also G. True experiment or fair test, because you'll keep all variables the same except the type of soil.

2. F. Design-redesign, because you'll test and evaluate different designs.

3. B. Observation, because you'll look, count, and record without changing the environment.

4. E. Screening, because you'll measure and compare the fat content of different brands.

5. G. True experiment or fair test, because you'll keep all conditions the same except the color of the food pellets.

6. D. Correlation, because you'll find and compare students' hand preference and language grades.

7. E. Screening, because you'll compare several different bags for their rotting times.

   Also G. True experiment or fair test, because you'll keep all variables the same except the kind of bag buried.

8. F. Design-redesign, because you'll write and test different programs.

9. E. Screening, because you'll compare how well different materials hold heat.

   Also G. True experiment or fair test, because you'll keep all variables the same except the kind of insulating material used.

10. C. Survey, because you'll ask people their ages and preferences.

    Also D. Correlation, because you'll look for a relationship between age and TV choices.

11. D. Correlation, because you'll see if students' grades in the two subjects change in the same way.

12. A. Model, because you'll show the arrangement of atoms in a molecule.

Set # 5, page 31

1. Weigh the ice cubes. The one that weighs the *least* melted the *most*. (They had to weigh the same at the start.)

2. If *best* means fewer peanuts, count the number of peanuts per 100 grams of mix.

3. While not revealing brands, ask people to rate the taste of samples on a point scale (perhaps 1 to 10).

4. Compare times in a 100-meter race.

5. Build identical parachutes of different materials. If *better* means stay in the air longer, record times until touchdown when dropping the parachute from a measured height.

6. If *better* means faster, record times to solve a problem or accomplish a task.

7. Count numbers of Canada geese at the pond every Friday at 4 P.M. from November through April.

8. If *better* means higher in the water, find apples and oranges of the *same weight*. Measure how low in the water (distance below the surface) they float.

9. Count the number of leaves, or measure the height of the plant or the circumference of its stem.

10. Count the number of ball games attended in a season or the number of games watched on TV by both teens and adults.

Set # 6, page 34

1. If light causes the pupil to constrict, then pupil diameter will be smaller in the light than in the dark.

2. If potting soil holds more water than sand, then a 100-gram sample of water-saturated potting soil will weigh less after it has dried than will a 100-gram sample of water-saturated sand after it has dried.

3. If rock music increases heart rate, then people's pulse counts for one minute will be higher immediately after hearing five minutes of rock music than before.

4. If girls remember numbers better than boys, then—after ten minutes—girls will be able to write down more of the numbers they heard on a taped message than will boys.

5. If country air has more dust than city air, then more particles will stick on a 25-square-centimeter greased card placed at O'Malley's farm than on an identical card placed at the corner of 5th and Vine. (The cards must stay in place for exactly the same amount of time.)

6. If seashells can purify water, then water filtered through seashells will have less lead than unfiltered water.

7. If pea plants lose more water in light than in darkness, then plants kept in the light for 24 hours will weigh less than identical plants kept in the dark for the same amount of time.

Set # 7, page 37

1. Rule #10. Put it off. You can always throw something together the night before it's due.

2. Rule #7. Ignore the rules.

3. Rule #9. Do your experiment only once or just ask a few people a couple of questions. Isn't that enough?

   Also Rule #3. Cut all the corners you can. Get by doing as little work as possible.

4. Rule #2. Write to experts asking for "everything you have on. . . ."

5. Rule #4. Quit the first time something goes wrong.

6. Rule #6. Make a pretty display. Who cares if it actually means anything?

7. Rule #1. Pick a boring topic you couldn't care less about.

8. Rule #5. Count on friends, parents, or teachers to bail you out at the last minute.

9. Rule #8. Don't bother to keep records or take pictures.

CHAPTER TWO

Fact Finding

Stop! Look! Listen!

By now you probably have a pretty good idea for a project. If all is well, you feel enthusiastic and eager to begin. Hang onto that enthusiasm, but slow down a minute. Before rushing into the laboratory or the field, you must do your homework.

Homework?

This homework isn't the kind that the teacher assigns but the kind you assign yourself. Homework for your project means mining the library, the community, and the Internet (if you can). You need to dig up those golden nuggets of information that separate good projects from great ones.

Ten (plus one) super reasons why fact finding makes sense:

1. The more you know, the less time you'll waste.
2. You won't reinvent the wheel, discovering what's already been discovered.
3. Ideas for improving your project will pop up where you least expected.
4. You'll find people you can ask for help.
5. The more you know, the more eager experts will be to help you.

6. You'll find people who want to know what you find out.
7. You won't be embarrassed when you ask for help.
8. By working from what's already known, you may be able to ask a truly new question and find a truly new answer.
9. The more you learn, the more confident you'll feel.
10. The more confident you feel, the more you'll impress teachers and judges.

Here's number #11, perhaps the most important of all:

11. The more confident you feel, the more fun you'll have!

PROSPECTING FOR INFORMATION

At the library

A good project starts first in your imagination. Your next stop? The library. Start at school, then visit the public library nearest you. Most college and university libraries are open to the community, too, so take advantage of that resource if you have one nearby. Also, don't overlook the libraries maintained by museums, civic organizations, community groups, and even political parties.

Close your eyes, and see yourself walking into the library. Does it feel a little scary—all those books and paper and computers everywhere? Where will you start?

Your library contains so much gold, you may need a treasure map.

Your search for gold will take you to several different places in that treasure trove that is your library. As you seek out prized nuggets of information, you will get more and more specific in your search.

ALERT: This book mentions only a few of the many resources your library has to offer. To dig deeper, read *Painless Research Papers* by Rebecca S. Elliot and James Elliot (Hauppauge, NY: Barron's Educational Series, Inc., 1998).

Start with the general. You need to understand your topic in its broadest sense before you begin. For example:

- If you want to research plant growth, start with a general knowledge of photosynthesis (how plants make food) and tropisms (the responses of plants to environmental factors such as light or gravity).
- If your project involves radiation, you'll need to understand what radiation is and how to detect and measure different kinds.
- If you want to build something using those "magic" plastic sheets that polarize light, then reading some general information about light will help you understand how light behaves and how polarizers work.

For background about your topic, the **encyclopedia** is a good place to begin. Any good, general encyclopedia—available either on the library shelf or on CD-ROM—will help get you started. Specialized volumes such as *Van Nostrand's Scientific Encyclopedia* and the *McGraw-Hill Encyclopedia of Science and Technology* may help, too, but they're only a beginning. You probably won't find anything specifically about your project idea.

So, where next?

Every library has a catalog of holdings: either a **card catalog** or a **computerized index**. Look under general topics such as *plants, radiation*, or *light*. You may find books related to your interest. You may also find useful videos, slide sets, and films. You may find one or more books that look interesting, but are you finished?

Nope. You're just getting started.

It's rare to find even one book that exactly fits your project. So, how do you get more specific? First, look in the **vertical file**. Librarians clip and save brochures and articles about a variety of topics. They often gather these in folders and store them in file cabinets. See what your library has filed on your topic. Then move on to **periodicals**.

Periodicals are newspapers, magazines, and journals published regularly—sometimes weekly, monthly, every two months, or quarterly. Some periodicals also put out special issues every few months or once a year. Periodicals range from covering very general interests and attracting general readership to publishing articles about highly specialized fields. For example:

- Major newspapers such as the *New York Times* and the *Wall Street Journal* often carry in-depth articles on science topics.
- Popular titles for general audiences include *Time, Newsweek, Life, U.S. News and World Report, Prevention,* and *American Health.*
- Reports about research in health care and news about environmental topics often appear in magazines that target female readers, such as *Good Housekeeping, Better Homes and Gardens, Family Circle, Woman's Day, Redbook,* and many others.
- Scientists often read general-interest journals such as *Science News,* the *Journal of the American Medical Association,* the *New England Journal of Medicine, Science, Scientific American, New Scientist,* or *Nature.*
- Every specialized field of science has its own association and publications, sometimes more than one. For example:
  - *Journal of Atmospheric Sciences, Journal of Physical Oceanography,* and *Monthly Weather Review*—all from the American Meteorological Society
  - *Wind Energy Weekly,* published by the American Wind Energy Association
  - American Mathematical Society's *Mathematical Reviews*
  - *Human Genetics* published by the American Society of Human Genetics
  - The American Association of Petroleum Geologists' publication *Explorer*

- People who share business, professional, political, or recreational interests form societies, associations, and clubs that often publish their own newsletters and journals. For example:
    - *Sea Notes* published by the Surfers' Environmental Alliance
    - *National Wildlife* and *International Wildlife* from the National Wildlife Federation

- *The NFPA Journal* of the National Fire Protection Association
- The International Association of Calculator Collectors publication *International Calculator Collector*
- *QST* from The American Radio Relay League (ham radio operators)

Name an interest, industry or avocation: There's probably a club, society, or association of people for it. Find an organization, and you'll tap into a supportive community of people who are just as excited about your topic as you are. You'll probably find a gold mine of information for your project, but beware of *bias*. Many groups have a political or social agenda.

How can you find societies, associations, and periodicals? The checklists "Looking for Help? Have You Considered These Experts?" on page 260 and "Organizations and Agencies" on page 265 may help you get started.

In addition, your library contains many user-friendly tools that will guide you in the right direction.

- Ask your librarian for indexes to newspapers such as the *New York Times*. Sometimes these are stored on microfiche (sheets of photographic film). Your librarian can show you how to use a reader with a screen to view them. Sometimes indexes are stored on computer disk or CD-ROM, which are faster and easier to use than microfiche.
- You'll have a hard time finding a library that doesn't have a series of volumes titled *Reader's Guide to Periodical Literature*. The books are easy to use. Check under your subject, and you'll find listings of articles in periodicals for the time period covered by the book. Larger libraries may have *Reader's Guide* on computer. Ask your librarian.

- Look for either printed or computerized versions of indexes to scientific and medical journals, including *Index Medicus*, *General Science Index*, or *Biological and Agricultural Index*.
- Look in the *Encyclopedia of Associations* or *Standard and Poor's Register of Corporations* (thick volumes shelved in the reference section). You'll probably find a business, foundation, alliance, or charity that shares your interest.
- Most larger libraries have some kind of periodical search capability on computer. Some commonly used search and retrieval systems include *InfoTrac*, *Proquest*, *SIRS*, *Dialog*, and others. For newspapers, try *NewsBank*—a selection of the best articles from newspapers over the past two or three years.
- If you have access to the Internet, a good place to begin looking for associations, special interest groups, and periodicals is the Internet Public Library at *http://www.ipl.org/*

Hold it!

You need a bit of savvy to understand the listing you find in indexes such as *Reader's Guide*. You won't find the article you need in indexes. Rather you'll find a *citation*. A citation includes the information you need to decide if an article might help and to find the complete article. For example,

Author **Year (month) of publication** **Title of article**

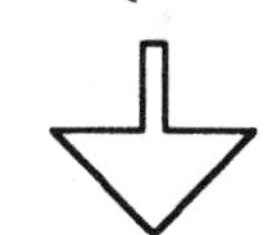

Abbot, Archibald. 1998 (October). The migratory habits of the blue-winged warbler *Vermivora pinus*. *Journal of the American Bird Watchers' Association* 44(10): 299–301.

The periodical's name, volume, number, and page numbers in which the article was published.

As you study a citation ask yourself:

- Who wrote the article? Have I found books or articles by this same author in other places?
- Is the article current enough to be useful? (An article from 1980 on acid-base chemistry might be fine; however, if you're looking for the newest advances in the manufacture of plastics or the mapping of genes, find a more recent article.)
- Is it specialized enough for what I want?
- Is it too specialized for me to read and understand?
- How will I use this information once I get it? (Having a clear idea of what you want and why will save you and your librarian a lot of time and effort.)

Problem: You find a citation to what sounds like the perfect book or article, but your library doesn't have it.

Solution: Most libraries can request books and articles from other libraries. The service is called an interlibrary loan. Allow plenty of time. Sometimes the search can take weeks. Sometimes the service is free, but sometimes you are charged. Ask ahead of time to make sure you can afford the cost.

On-line

If your library is a gold mine of knowledge, then the Internet is an ocean of information. The Internet can be the science student's best friend, but don't waste time surfing the backwaters.

You can ride the crest of the wave if you:

- Learn what search engines are and how to use them. Ask your teacher or librarian to get you started with *Infoseek, Yahoo!, Alta Vista, Hotbot, Excite, Lycos*, or any of the other popular search engines. *Read the instructions*. Each search engine has its own special techniques for helping you get to specifically what you want without getting a thousand things you don't want.
- Follow links. The links at one Internet site can often lead you to another that's just what you need. Note the addresses of the useful sites you find. The address is the site's URL (Uniform Resource Locator), and you'll need it to return to that page later. Most browsers permit saving URL's in a list of "Bookmarks" or "Favorites." The site's URL will also appear on pages you download or print.

Caution—Major Mistake Territory!

Consider the Source. If the site's URL contains .gov (for government) or .edu (for an educational institution), then the information *should be* accurate and the file virus free, but no guarantees exist. Entries containing .com are businesses or industries, and they may be trying to sell something. URLs containing .org (for organization) may lead you to associations that have a particular mission, so look out for bias. Approach most cautiously entries in newsgroups and chatrooms or items on personal pages. Far-out, wild, and zany pages may be fun, but risky, sources. Download, save, or print only the best.

- Save those files. Find out how to download material from the Internet to a floppy disk. You can also ask your librarian for permission to print a copy. Don't spend hours searching out valuable information only to find that you can't retrieve it later.
- Use the Net to make contacts. If pages include e-mail addresses of experts willing to answer questions, write to them. However, make sure you know your subject well and can ask a specific question before you punch that send button.
- Some libraries also subscribe to information retrieval banks that charge for articles. If you do a lot of research for school, you may even want to subscribe on your computer at home. Two excellent, relatively inexpensive services are Electric Library (*http://www.elibrary.com*) and Northern Light (*http://www.northernlight.com*). Also, *Science Online* may be available in larger libraries that subscribe to *Science*, the journal of the American Association for the Advancement of Science.

- If you need (and feel you can handle) highly technical reports, try one of the on-line search engines serving specialists in a field. For example, the National Center for Biotechnology Information and the National Library of Medicine offer an easy-to-use search engine at *http://www.meds.com.*

CITING YOUR SOURCES

It's not enough to find an article that will help you with your project. You will need to let everyone else know what it is and how to find it. Some students use index cards to keep track of sources. Others prefer to make photocopies of articles and write source information directly on the article. Either way, make sure you keep track of (at a minimum) the following pieces of information about the source.

For articles:

- author's or authors' name(s);
- title of the article;
- periodical in which you found the article;
- date of the periodical, volume, and issue numbers (usually printed on the cover or on the table of contents page); and
- page numbers of the article.

For newspapers:

- author's or authors' name(s);
- title of the article;
- name of the newspaper;
- date of publication; and
- section and page numbers.

For books:

- author's or authors' name(s);
- title of the book;
- name of the publisher;
- city, state, and country where the publisher is located;
- copyright date (look on the back of the title page for the latest year following the symbol ©); and
- numbers of the pages read or quoted.

For on-line sources:

- author (if given);
- title of the document;
- name of the organization, educational institution, or company that sponsored the posting;
- place where the sponsor is located;
- date given on posting or page; and
- URL (Uniform Resource Locator), that is, the document's on-line address.

Your teacher will tell you what form to use in your citations. Follow it exactly. Be especially careful about capitalization and punctuation. Here are two standard forms of citation often used for articles:

SCIENTIFIC ("DOWN" STYLE):

Smith, Joseph. 1998 (April). Development of K-12 instrumentation for sub-alluvial infiltration. *Mining and hard rock news* 16(4): 23–26.

CLASSIC OR LITERARY STYLE:

Joseph Smith. "Development of K-12 Instrumentation for Sub-alluvial Instrumentation," *Mining and Hard Rock News,* April 1998, 23–26.

BRAIN TICKLERS
Set # 8

Jasmine's project on color blindness got a real boost in the library. She found a lot of useful material because she knew where to look—and how. Try to match her savvy by matching the following items with the places Jasmine found them.

Item	Source
1. Definition of *monochromatism*	a. *Reader's Guide* or computerized index of popular periodicals
2. "Numbers and Ratios of Visual Pigment Genes for Normal Red-Green Color Vision" by M. Neitz and J. Neitz, *Science* (February 17, 1995), pgs. 1013–1016.	b. The card catalog or computerized index of library holdings
3. Definitions of different forms of color blindness.	c. Medline search at Internet address *http://www.meds.com*
4. Video: *How Animals See Colors*	d. Internet search at *http://www.solarchromic.com*
5. Pamphlet: *Inherited Eye Diseases* (by Prevent Blindness America, Schaumberg, IL)	e. *Dictionary of Modern Biology* by Norah Rudin (Barron's, 1997)

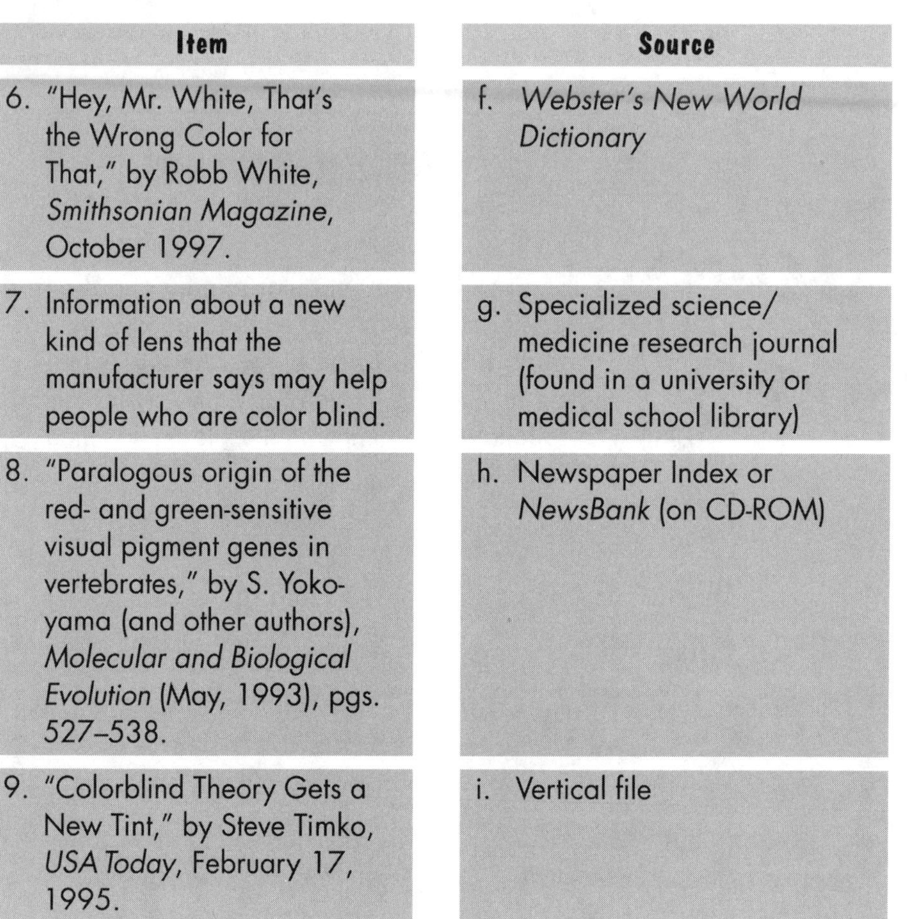

Item	Source
6. "Hey, Mr. White, That's the Wrong Color for That," by Robb White, *Smithsonian Magazine*, October 1997.	f. *Webster's New World Dictionary*
7. Information about a new kind of lens that the manufacturer says may help people who are color blind.	g. Specialized science/ medicine research journal (found in a university or medical school library)
8. "Paralogous origin of the red- and green-sensitive visual pigment genes in vertebrates," by S. Yokoyama (and other authors), *Molecular and Biological Evolution* (May, 1993), pgs. 527–538.	h. Newspaper Index or *NewsBank* (on CD-ROM)
9. "Colorblind Theory Gets a New Tint," by Steve Timko, *USA Today*, February 17, 1995.	i. Vertical file

(Answers are on page 82.)

HOW USEFUL IS IT?

Okay. You've found an article about your topic. Does that mean it's

- timely?
- true?
- complete?

- factual?
- unbiased?
- useful?

Maybe.
Here's how to decide:

1. *Make sure it's recent.* Check the copyright date on books. Check the publication date on periodicals and articles. Material that is more than five years old is always suspect. Even two- or three-year-old information can be old stuff in rapidly changing fields of science and technology. Use only the most recent sources available.
2. *Consider the source.* Information from the school's astronomy club may not be as trustworthy as official publications of the American Astronomical Society. Kenny Conman's personal Web site may be fun, but material from official organizations, educational institutions, and government publications is probably more reliable.
3. *Double- and triple-check facts.* Don't trust a single source of information. If it's true, it will be reported in more than one place. Dig, dig, dig to confirm your facts just as reporters and journalists do.
4. *Never stop digging.* You will never learn it all. Keep looking before, during, and after you complete your project.
5. *Select facts carefully.* You can easily copy everything you find, but what's the use? It is better to pick and choose the facts and ideas that suit your project perfectly. No magic formula will tell you how to judge relevance. Making that most difficult of all decisions—what to leave *out*—requires thought, thought, and more thought.
6. *Identify bias, and turn it to your advantage.* Biased material need not go out in the trash. It may lend depth and substance to your project *if* you can counter it with arguments from the other side of a controversy. For example, an environmental protection society may oppose the construction of a new highway, while the local Chamber of Commerce may favor it. Take either of these points of view in isolation and you'll become a victim of bias. If

you look at the two together, however, you may well end up with a more or less complete picture of the issue and the options.

7. *Don't confuse facts, opinions, and persuasive arguments.* Although opinions and persuasive arguments may add interest to your project, it's "just the facts ma'am" that provide the sound basis you need for your work.

How can you distinguish fact, opinion, and persuasion?

FACT:

A truth that can be recorded, documented, observed, proved, or replicated in a way that all unbiased observers can agree on its objective reality. For example:

- On Saturday, May 12, the sun shone at Denver International Airport for six hours and 43 minutes without interruption.
- In the Northern Hemisphere, the average length of daylight in a 24-hour period is longer in July than in December.

- In a trial involving 50 volunteers conducted at Denver General Hospital, 74 percent of those tested showed a reduced rate of hair loss after six months of using No-Go-Bald scalp lotion.

OPINION:

A personal point of view, judgment, or belief. For example:

- Dr. Fenton Floss, a dentist with an office in North Highland Shopping Center, says most kids could be cavity free if they'd quit eating candy, cake, and soda. (That's Dr. Floss's opinion. Does any evidence support his view?)
- Thirty-seven members of Bigfork Athletic Club say that an extra dose of vitamins A, C, and E gives them more energy for their workouts. (How many disagree? How many don't know? Have any research studies supported or refuted this idea?)
- No-head-rin is the best pill to take for headaches. How do I know? I use it. (Maybe you do, but does that mean it is "best" by some objective measure?)
- The government sets so-called "safe" levels for pollutants. We say the only safe level is zero. (What pollutant? What is the government's safe level? How is it set? Is a zero level attainable?)
- Popsicle Popover is the best band in Meterville. Everybody knows that. ("Best" by what measure? Who's "everybody?")

PERSUASIVE ARGUMENT:

An opinion offered to convince others to take a particular course of action. For example:

- Drink Lemon-LuLu Soda. It's tangier, zippier, full of flavor and fun. (Tangier and zippier than what? Who measures tanginess and zippiness, and how? How can one soda be more fun than another? Isn't the mission here to sell soda?)
- Elect Phineas Fowl and watch your taxes soar sky high. (This is what Fowl's opponents say. What evidence supports their view? What does Fowl himself say on the subject? Does he offer any convincing evidence to the contrary?)
- All your friends have seen it! Don't be a nerd and miss it! *Brain Dead* is the movie of the year. (Have all your friends truly seen it? So what if they have? Will missing it really make you a nerd? What does "movie of the year" mean? The best attended? The biggest moneymaker? Are the producers the advertising agency's best client?)
- NFL quarterbacks shower with Zippy-Suds soap. Shouldn't you? (Do all of them really use it? Even if many of them do, so what? It could cause you to break out in hives.)

BRAIN TICKLERS
Set # 9

Savvy science projects use facts, facts, and more facts, and stay away from opinions and persuasion. Label each of the following as F (fact), O (opinion), or P (persuasion). Explain the reason for your label.

Hints: (1) Not sure if it's a fact? Look it up! (2) Sometimes opinions and persuasion seem a lot alike. Opinion becomes persuasion when it seeks to change the thoughts or actions of someone other than the opinion holder.

1. Rapid burial of hard (skeletal) body parts provides the most favorable conditions for fossilization.
2. Fossil hunters are more interested in finding treasure than in advancing science.
3. If you want to get rich and famous, the last thing to do is go fossil hunting.
4. Hold a crystal in your hand, and your cold will pass in hours instead of days.
5. I always wear a crystal to ward off colds.
6. Certain crystals with no center of symmetry exhibit the piezoelectric effect.
7. Pig iron is used to make steel.
8. Investing in the steel industry is a sure way to lose money these days.

9. If you want to make a killing in the market, pick up some penny stocks in the steel industry.
10. I don't know, but I've been told the streets of York are paved with gold.

(Answers are on page 83.)

RECORD KEEPING

Whatever your project, you must keep detailed and accurate records from beginning to end.

Your journal

If you have not already done so, purchase a bound notebook. (You can easily lose things from a loose-leaf notebook.) This book is your journal. Treat it as a diary of your project. Record in it everything you think about and do day by day to move your project along. Nobody sees your journal but you, so leave nothing out.

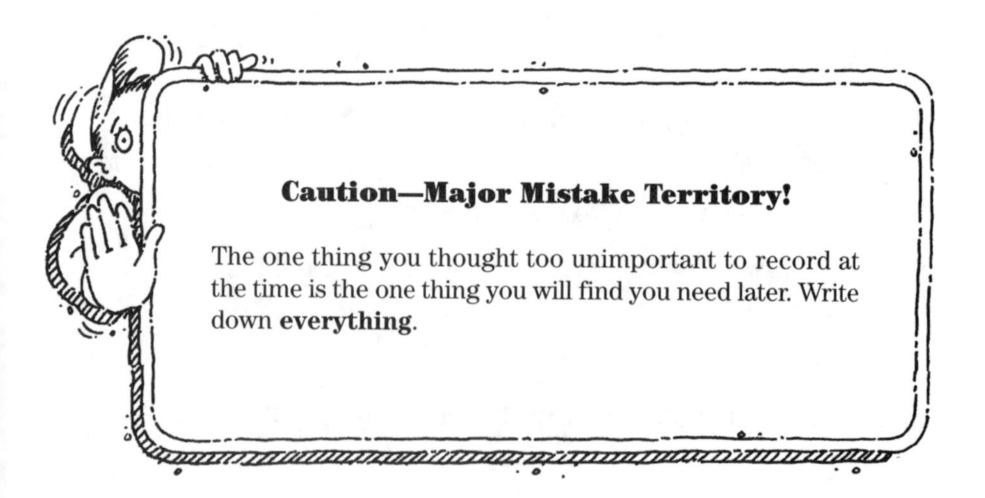

What should you put into your journal?

- Your personal copy of "Do-It-Yourself Plan and Schedule" from page 247
- Your personal copy of "What Am I Learning? Watch Your Skills and Knowledge Grow" from page 249
- Any ideas that pop into your head
- Conversations you have with people (record their names, addresses, and phone numbers)
- Places you visit (when and why)
- Letters you write or receive
- Citations and notes from your library and on-line research
- The design of your experiment
- Experimental results: numbers, numbers, numbers
- The things you do for your project
- Problems and how you solve them
- Feelings about your work (yes, it's OK to write down your frustrations)
- Anything and everything—and include the date—because memories are less than perfect

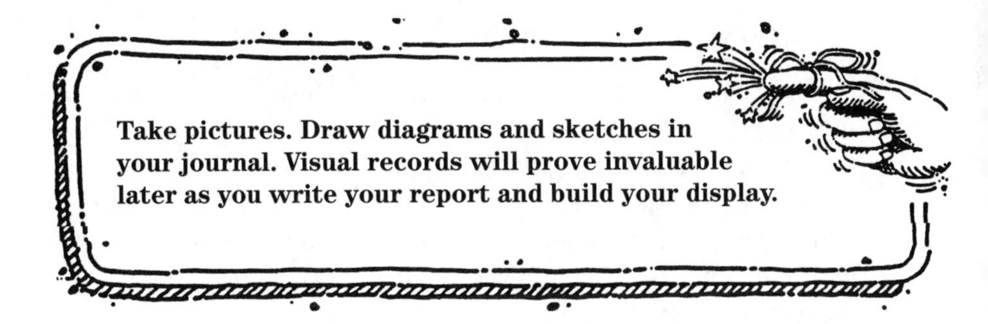

Your logbook

When your project is nearing completion, you will want to transfer some of your records to another notebook, your logbook. You will show your logbook to others as part of your display. Copy into it a carefully chosen selection of entries from your journal. What goes into your logbook?

- Descriptions and data from your experiments
- Library references
- Sketches of experimental designs
- Photographs
- Sources of help and information

Everything in your logbook must be in order, dated, neat, complete, and accurate. Leave everything else in your journal for future reference.

GETTING HELP

Part of fact finding is getting help from people. You may already have begun to wonder

- What kind of help do I need?
- Who can help me?
- How much help is too little?
- How much help is too much?

These are tough questions to answer. The amounts and kinds of help you seek will depend on the nature of your project, the help available, and your own plans for your work. Just remember, this project is yours, not anyone else's. Some well-meaning adults may unintentionally try to take over your project. Don't let it happen. Seek the help you need and use it wisely, but remember the following tip.

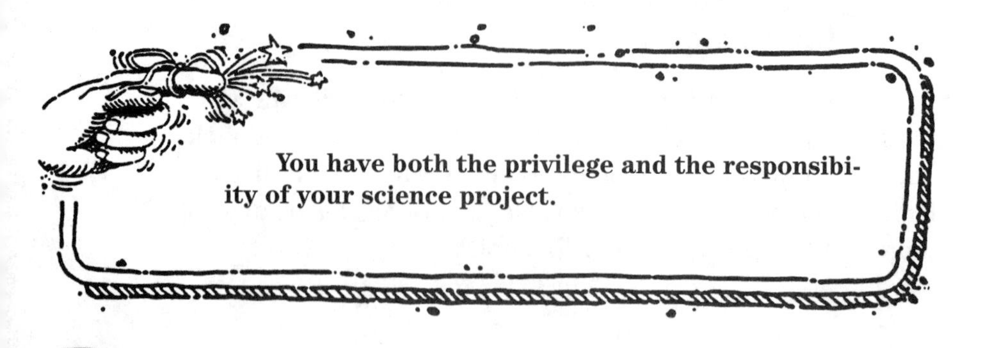

You have both the privilege and the responsibiity of your science project.

Think about the lessons in this true story. Louise wanted to build a model of a motion picture machine. She designed a holder for a revolving cake pan. She needed a pyramid with a spike protruding from the top, she decided. Neither she nor her mother had any power tools at home, and neither knew how to construct the device. "Have you asked the industrial arts teacher for help?" her mother asked.

Louise was delighted when the teacher took her into the shop and showed her how to shape the wood and insert the spike. Louise tried three times before she got the size and shape she wanted. In the end, though, she won first place for her grade in the school science fair. Louise went on to study science in college. Today she writes science books and articles for teens and adults.

What did Louise do right?

- She knew what she wanted and needed for her project. She didn't ask someone else to design her model for her.
- She realized the value of her mother's suggestion but acted on it herself. Her mother didn't ask the industrial arts teacher for help. Louise did.
- Louise built her device herself. The teacher helped her learn new skills, but he didn't do the work for her.
- Louise kept at it. When her first attempt failed, she tried again.
- Louise didn't need to win that fair. She would have chosen a career in science anyway because, from this project, she found she loved the work.

You can follow Louise's example by finding the help you need and using it wisely. As in Louise's story, parents, friends, teachers, and neighbors can often be good sources of help. They may offer ideas to help you get started or technical assistance along the way.

Expert help is available in many communities, often from sources you may not think about at first. (See "Looking for Help? Have You Considered These Experts?" on page 260.) Also, some schools offer programs through which students can team up with mentors. The best mentors act as coaches. They can show you and tell you how to play the game, but putting points on the scoreboard is up to you.

Never overlook the value of a good critic. Ask adults you respect to look at your work and tell you what they think is wrong with it. Develop a thick skin. They aren't criticizing you. They are suggesting ways to make your work better. You don't have to act on everything somebody tells you, but your project will be all the better if you consider many different points of view.

Attention Parents, Teachers, and Mentors: The checklist "How Can I Help? A Checklist for Parents, Teachers, and Mentors" on page 264 suggests some positive ways that you can assist students with their science projects.

RULE OF THUMB:

You will be better off having too little help than too much. You can always dream up ways to make your project better all on your own—without anyone's help; however, you'll have a hard time regaining control if you've let someone slip into the driver's seat.

Caution—Major Mistake Territory!

Think of your project as an airplane and yourself as its pilot. Copilots give advice, but they don't grab the controls or reverse the course the pilot has set. Have confidence in yourself to know how much help you truly need and when. Never doubt that you can fly your craft like a pro and come in for a safe landing on home field.

TRAPS AND TRIP WIRES: COMMON PROBLEMS TO AVOID

You now stand at a turning point. You know a lot about your subject, and you are beginning to have a pretty good idea of what your project can become. Before you go too far, however, ask yourself some questions. The answers may help you avoid the problems that most often trap or trip even the most dedicated science student.

1. *Can I afford this project?* Most science projects require some investment in materials, supplies, and equipment. If the cost of your project exceeds your allowance, you may find yourself bankrupt before you ever start. Ask yourself if you can cut costs by substituting cheaper materials or borrowing rather than buying. If finances are seriously

impeding you, discuss the problem in private with your teacher. Scholarship or school monies may be available to assist students.

2. *Do I have enough time?* Too many students make the mistake of thinking they have to start and complete a new project every year. In truth, you may be better off continuing your research from one year to the next—even taking an idea from middle school on into high school. Sticking with a project over two, three, even four years lets you delve deeply into a question. One student who stuck with a topic found himself mentioned in the *Los Angeles Times*. He spent two years studying some corn he grew in his backyard.

For major, prestigious competitions, your work must continue over several years. Check the newspapers and science magazines late each spring, and read about the winners of the Westinghouse Science Talent Search. These high school seniors—who can win scholarships as generous as $40,000— usually present projects they've taken three or four years to complete. Sometimes, their prize-winning research grows out of projects started in middle school or even in elementary school.

3. *Can I get the equipment and supplies I need?* Good carpenters check their tools before they begin building. You should do the same. You don't need to own that special piece of equipment or even have it in your school, but you do need to make sure you can use it in someone's lab or borrow it from some source. Don't be discouraged if that special thing you need sounds unattainable. You might be surprised what you can order from a supply house (for a small investment) or locate in your community. The checklists "Supplies and Materials You Thought

You Couldn't Get" on page 256 and "Scientific Supply Houses" on page 270 may help you get started.

4. *Do I need to use hazardous chemicals or equipment?* If so, check with your teacher. Science fairs usually ban potentially dangerous substances and devices, but may allow some with permission. (See "Safety First" on page 251 and "Do Not Display" on page 262.)
5. *Do I need permission to use animal or human subjects?* When in doubt, ask. When not in doubt, ask just the same.
6. *Am I relying too much on others?* To a science fair judge or experienced teacher, that project your mother or older brother did for you (to help you out of a jam, of course) stands out like a pumpkin in a cabbage patch. Maybe your Uncle Larry or your Cousin Sabrina can do a great project, but that's their project, not yours. Keep control over your own work. Don't let an enthusiastic, well-meaning friend take your project away from you. Remove any doubt from your own mind, your teacher's, and the judges'. Keep detailed records in your journal so that your final logbook and report can reveal exactly who helped you and how.
7. *Does the thought of competition make me shiver?* Some people thrive on competition. Others would rather hike the Yukon barefoot than compete with anyone in anything. Depending on the rules of your science project, you may find yourself vying against others—whether you like it or not. What should you do if you're the frigid-footed type? Don't think about being in a contest with

other students. Make this, instead, a contest with yourself. Do what athletes do: strive for your personal best. Remember, it's not the destination that matters, but the journey. The **process** of completing your project is more important than the **outcome**.

8. *Can I keep myself going during the long period of time it takes to complete a science project?* Too many students begin with vigor but run out of steam as time goes on. Keep yourself on track by choosing a project that interests you and by remembering your goal. Again,

take a tip from athletes. They visualize themselves performing at their best—seeing in their mind's eye that winning high jump or hurdle race. You can do the same. See yourself presenting a project about which you feel proud. Imagine that good feeling that comes from completing a piece of work you know is good.

A little reminder never hurts either. Find a motivating quote—one of the following or one of your own—and copy it onto index cards. Tape a card to your mirror. Carry one in your pocket. Sleep with one under your pillow. It works for athletes. It can work for you.

"You can run with the big dogs or sit on the porch and bark."
—Wallace Arnold

"The important thing is not to stop questioning."
—Albert Einstein

"Seventy percent of success in life is showing up."
—Woody Allen

"Research is what I'm doing when I don't know what I'm doing."
—Werner von Braun

"Nothing in life is to be feared. It is only to be understood."
—Marie Curie

"When you reach for the stars, you may not quite get one, but you won't come up with a handful of mud either."
—Leo Burnett

BRAIN TICKLERS—THE ANSWERS

Set # 8, page 65

1. f. The dictionary is the best place to look for single word definitions.
2. g. Many public libraries subscribe to *Science*, too.
3. e. Dictionaries of science, technology, and medicine offer more detail than general dictionaries.
4. b. Your library probably has more than books. Look for videos, sound recordings, CD-ROMs, and other useful tools.
5. i. Find pamphlets in the vertical file or request them from associations and societies.
6. a. High quality, popular periodicals are a great source for easy-to-read and up-to-date information.

7. d. The key word here is *manufacturer*. Companies often publicize their new products via the Internet. Notice the .com ending of the URL.

8. c. Not for the faint hearted—consult specialized journals once you become an expert on your topic.

9. h. If it's new, it's in the newspaper. Consult a newspaper index as well as CD-ROM resources such as *NewsBank*.

Set # 9, page 71

1. **F**

2. **O**. This sounds like nothing more than the speaker's uninformed prejudice.

3. **P**. You should have a measure of healthy skepticism here. You should wonder why this person wants to keep you away from the dinosaur bones.

4. **P.** No medical evidence supports this claim. You can try it if you like, but if your cold passes, how will you know the crystal made a difference? Maybe your cold would have passed quickly without it.

5. **O.** Go ahead if you like. What harm can it do? However, does any evidence show that it works?

6. **F**

7. **F**

8. **O.** Maybe the speaker has lost money. Maybe many others have lost money. However, does "sure" mean 100 percent? Check out the odds before agreeing or disagreeing.

9. **P.** Don't spend a penny until you have *all* the facts.

10. **F.** Surprise, surprise. This statement is a *fact*. The speaker admits to repeating mere hearsay. If you are interested in York's pavement materials, find some factual sources.

CHAPTER THREE
The Fair Test
Turkey Eating Contest
1

Now that you've become an expert on your topic and skeptical of unsupported claims, it's time to get down to business in planning your project. Your question should already have given you a good idea, but you'll want to make sure that any testing you do is **fair**. Scientists use some very strict rules for deciding what's fair and what isn't, but the commonsense meaning of the term is a good place to start.

What does *fair* mean to you?

- An even chance of winning?
- No one getting an advantage?
- Not playing favorites?
- Enforcing the same rules for all?

A fair test in science means all that and more. What unfair conditions can you spot in these project plans—all undertaken by students who wanted to find out whether sand or potting soil gets hotter in the sun?

- Hailey filled a frying pan with potting soil and a loaf pan with sand. She added thermometers and set her pans in the sun. (Not fair! The containers differ.)

- Kyle weighed out 100 grams of sand and 50 grams of potting soil. He put each sample into a separate, identical glass jar. (Not fair! The containers are the same, but the sample sizes differ.)
- Before beginning her experiment, Brianna knew she should check her thermometers. She put them into warm water. One read 42°C. The other read 38°C. "Close enough," she decided and used one in sand, the other in potting soil. (Not fair! The thermometers should read the same before the experiment begins.)
- Kerry knew he should repeat his experiment several times and average his results. He ran out of potting soil after his first trial, so he used soil that he dug from his garden in his second and third trials. (Not fair! The soil type should be the same for all trials.)
- Mikayla's sand was dry in her first trial. However, she stored it outside, and it was wet when she conducted trials two and three. (Not fair! The characteristics of the soil should not change from one trial to another.)
- Chris put his sand in the sun and his potting soil under a desk lamp. (Not fair! Two different heat sources don't compare.)

- Troy put a thermometer into an oatmeal box, added sand, and set the box in the sun. He then repeated the same steps with another oatmeal box and potting soil. After an hour, he recorded the temperature in the boxes. (Not fair! The first box was in the sun longer than the second.)

VARIABLES AND CONTROLS

Each of the experiments described above was flawed because the student didn't *control* a *variable*:

- **Variable:** something that can change or vary
- **Controlled:** held constant; kept the same

In experimental projects, a variable can change for one of only two reasons:

1. You change it on purpose (the cause). This is your **independent variable.**
2. It changes as a *result* of the variable you changed on purpose. This *effect* or *outcome* is your **dependent variable**.

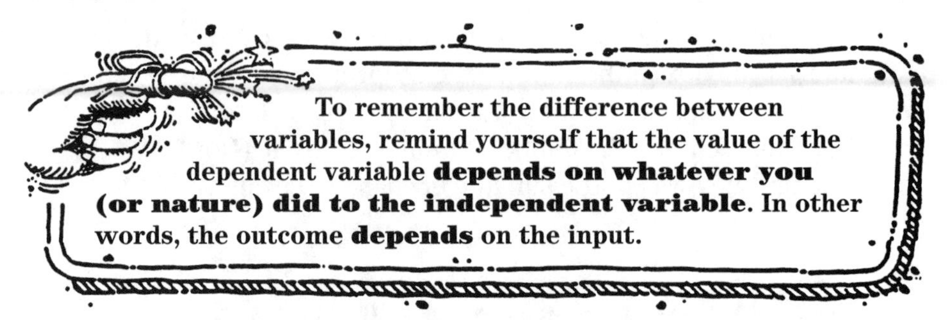

For those students trying to find out if potting soil or sand gets hotter in the sun, the type of soil is the independent variable. The temperature after exposure to light is the dependent variable.

A fair test means that no variable other than the independent and the dependent can change.

In the previous example of soil temperatures,

- all the containers must be the same size, shape, and material;
- the time of exposure to sunlight must be the same for all samples;
- the type and condition of the soil samples cannot vary from one trial to the next;
- samples sizes must be identical;
- measuring instruments (the thermometers) must read identically under identical conditions; and
- the light source must be the same for all trials.

All variables are potential sources of error. You must control them.

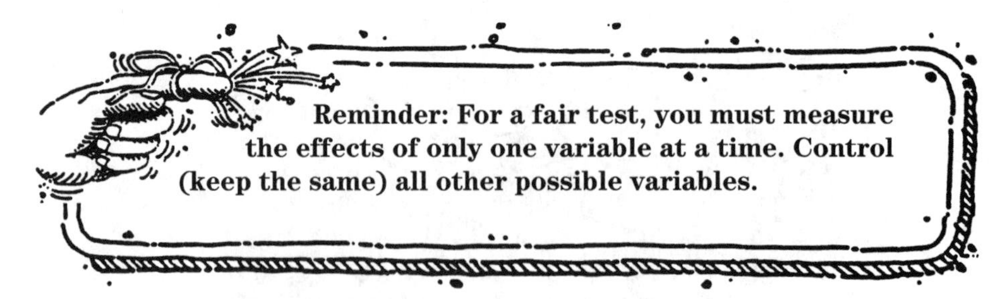

Before you go on, make sure you understand independent and dependent variables and control of variables. These examples may help:

Project title	Independent variable	Dependent variable	Other variables that must be controlled
What is the relationship between the length of a lever and the weight it can lift?	Length of the lever (measured from its support, the fulcrum)	Weight lifted	Size and material of the fulcrum; length, width, and material from which the lever is made; shape and material of the weights; how the weights are placed on the lever, and others.
Do people blink more in bright light or dim light?	Intensity of light (from same source ideally)	Number of eye blinks per minute in volunteer subjects	Sex, age, size, and other characteristics of subjects themselves; place where test is conducted—all aspects including temperature, wind, noise, other light sources; stopwatch (to ensure that all subjects are tested for the same length of time); instructions to subjects; time of day; and many others.
How do leaves vary in their need for water?	Type (species) of leaf	Amount of water taken up	Containers that hold the leaves; amount of water at beginning of experiment; time before measuring water uptake; size of leaf (must be the same or controlled mathematically); temperature, wind, light exposure, and all other environmental variables.
Which end of the egg can support more weight: the round end or the pointed one?	The end of the egg	The amount of weight supported	The system for testing and measuring, plus all environmental variables.

BRAIN TICKLERS
Set # 10

Identify the dependent and independent variables in each of the following experiments. Suggest at least one other variable that must be controlled to ensure a fair test. Use I to indicate the independent variable, D to indicate the dependent variable, and C to indicate a variable that must be controlled.

1. How does the amount of glycerin added to the bubble mix affect the length of time a bubble will last?
2. How much salt can water dissolve at different temperatures?
3. Can volunteers tell the difference between diet and regular cola?
4. Which plants grow new stems after being cut?
5. Do lightbulbs last longer when wired in parallel or in series?
6. Do ball bearings fall faster in corn syrup or glycerin?
7. What's the best way to keep bacteria from growing in gym shoes?
8. Do children sleep less than the elderly?
9. Do phases of the moon affect barometric pressure?
10. Can wastewater from the shower be used to water the lawn?

(Answers are on page 129.)

BIAS

Bias is intellectual blindness: you see only what you expect to see.

If you feel too sure of your outcome, you may blind yourself to a fair test. Convinced that you know the answer before you begin, you unconsciously see only those data that confirm your predetermined conclusion. Break the bias bind. Showing a hypothesis to be false can be just as important as showing that it is true.

Avoiding bias means making sure every test is fair and that your mind is open to all ideas. For example, can you spot the biases in Brad's project? How did Tanesha conduct the same project in an unbiased way?

BIASED:

Brad can't wait until he's old enough to drive a Yankee pickup truck. He's convinced that Yankees are the best on the road. For his project, he asks ten Yankee truck owners

how satisfied they are with gas mileage, acceleration, and handling (on a scale of 1–10). When he averages their answers, he gets ratings of 7 (mileage), 8 (acceleration), and 6 (handling). He concludes that Yankees are the best trucks on the road.

UNBIASED:

Tanesha knows she can't compare all trucks on all measures, so she chooses three she thinks are important: gas mileage, acceleration, and handling. She interviews 10 owners of each of six different makes (manufacturers) of trucks, asking each to rate the three variables on a scale of 1 to 10. When she averages and graphs her numbers, she finds that Yankee owners rate their vehicles higher than the other brands on handling but lower than the others on acceleration. Other makes and models also score higher on gas mileage. She compares the owners' ratings with facts and figures obtained from truck dealers. Tanesha finds that owners' ratings of gas mileage don't match the dealers' claims. She concludes that no single manufacturer can be shown to produce the pickup with the "best" combination of mileage, acceleration, and handling. Tanesha notes that many other indicators of owner satisfaction should be measured in future studies. These include number of repairs in the first six months, start-ups in cold weather, or resale value after two years. She needs some controlled, objective measures of gas mileage, too, she says.

Brad's biased project	Tanesha's unbiased project
Assumed that Yankees were better	Assumed that Yankees needed to be compared to other makes
Used data only from Yankee owners	Used data from owners of different makes
Used too few subjects	Used a greater number of subjects

Assumed that gas mileage, acceleration, and handling were the only important measures of owner satisfaction	Acknowledged that gas mileage, acceleration, and handling were only three among many measures of owner satisfaction
Used a single source of data (owners' answers)	Used more than one data source (owners' answers and information from dealers)
Drew a false conclusion	Drew conclusions supported by the data
Overgeneralized	Recognized the limits of the study and its outcomes

BRAIN TICKLERS
Set # 11

Correct the bias in each of these projects. Hint: You may need to correct more than one thing.

1. Stacey thinks 2-percent milk tastes better than full-fat or skim milk. She leaves samples of skim milk and full-fat milk in the refrigerator for a month. They both get spoiled and foul smelling, so Stacey knows she must be right.
2. It always rains more in April than any other time of the year. Hobart measures April's rainfall and shows that's true.
3. Kelly proves Brand X ketchup is the best by timing how long it takes a tablespoon of the red stuff to run down a board.

4. Coffman shows that Vital-Grow fertilizer is everything it's claimed to be by growing tomatoes a meter tall in less than two months.

5. Euralia proves that kids misbehave more when the weather is cold by reporting the number of detention hall slips teachers wrote in December, January, and February.

6. To prove that acid rain kills pond life, Monty collects samples from Hill Creek Pond where the pH measures 4.5 (very acidic) and draws pictures of the organisms he finds there.

7. Paige proves that Sunset Shampoo is good stuff by asking three friends to try it and tell her how they like it.
8. Dion extracts pigments from maple leaves in October. After finding no green pigment, he reports that chlorophyll disappears from trees when the weather turns cold.
9. Samantha watches three chameleons for a week and keeps a careful record of her observations. She decides they turn green when they're happy.
10. Martin offers his pet tortoise green lettuce, a red apple, and blueberries for food. From his pet's choice, Martin concludes that tortoises like red better than green or blue.

(Answers are on page 131.)

WAYS TO MEASURE

In science, it's not enough to say that things differ from one time to another. If something changes, the change must be measured.

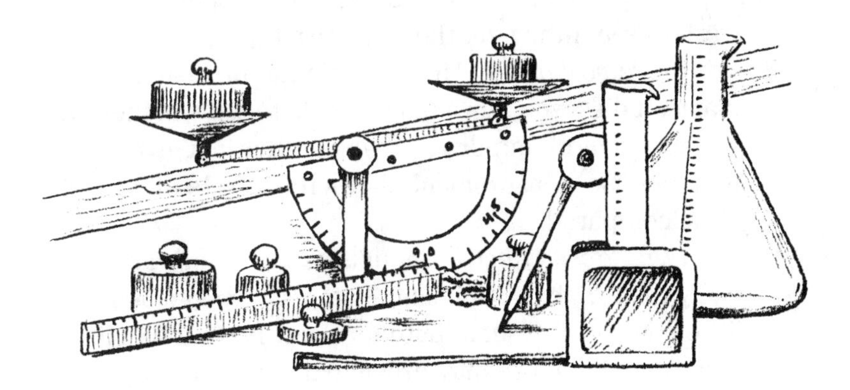

Ten reasons to measure **everything**:

1. *Control variables.* The fair test requires all variables except the independent and dependent to be held constant. The only way to make sure they stay the same is to measure them.
2. *Describe experimental conditions.* It's not enough to say that you changed the independent variable. The procedure for the experiment must answer the questions, "In what ways did you change it?" and "By how much?"
3. *Describe outcomes.* The same questions must be answered about the dependent variable, "In how many ways did it change?" and "By how much?"
4. *Determine significance.* Dependent variables can change just by chance. For change to be important, it must be too big to be the product of chance. (Probe the question of significance in Chapter Four.)
5. *Find Trends.* Measurements reveal if variables change in an orderly, systematic fashion.
6. *Predict.* Measurements may allow researchers to predict the value of a variable under conditions not actually measured.
7. *Replicate.* Scientists must be able to repeat each other's work to see if they get the same results.
8. *Compare variables.* After combining the results of many studies, scientists may be able to tell whether one variable is more important than another in a given system. Precise measurement allows the effects of variables to be compared.
9. *Find the "dose effect."* Amounts can be important, especially for variables such as pollutants or medicines. A little may not do much, but more may make a big difference, and still more may prove toxic or fatal.
10. *Communicate.* Scientists share their findings using agreed-upon systems of measurement.

Kinds of measurement

In general, scientists measure two categories of variables. **Objective** variables actually exist in the physical world (for example, length, weight, distance, height, temperature, and time). **Subjective** or "human" variables aren't so easily measured (such as people's behavior, preferences, choices, moods, or motives).

Objective variables can be measured with tools such as meter sticks and thermometers. Subjective variables require creative approaches to the construction and use of measuring devices.

Measuring objective variables

Review your project design. What objective variables do you need to measure? Height? Volume? Area? Time?

What about using the metric system?

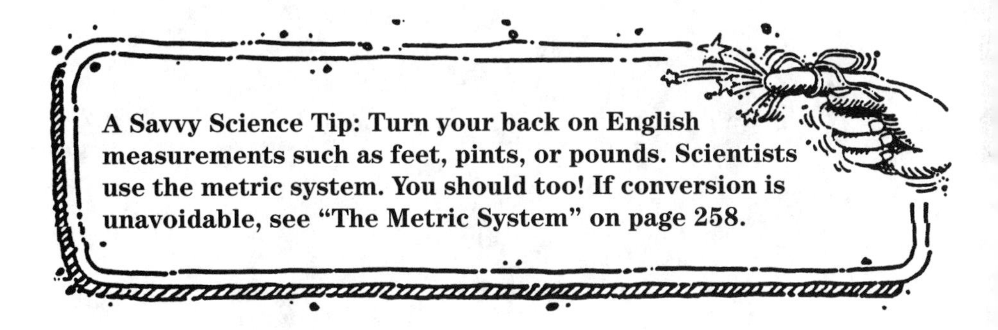

Here are some of the physical variables often measured in science projects and the units and devices used to measure them.

Variable	Unit	Measuring device	Notes
Length (width, height, depth, distance, or any other single dimension)	Meter (centimeter, millimeter)	Meter stick or measuring tape	A kilometer (1,000 meters) is a convenient unit to measure long distances.
Area (width x height) or the space enclosed by any two dimensions. For example, or	Square meter or square centimeter	Meter stick or measuring tape	Square kilometers or hectares are convenient units to measure large areas of land.

Volume (the amount of three-dimensional space an object occupies). For example, or	Cubic meter, liter, cubic centimeter, or milliliter	Meter stick or measuring tape (for straight-sided objects); graduated cylinder or measuring cup (for liquids and irregularly shaped objects) Note: Laboratory beakers and flasks are not marked accurately enough to measure with precision.	Partially fill a graduated cylinder or measuring cup with water. Record the water level. Add an object that sinks. For an object that floats, lightly push it just below the water's surface with the tip of a pencil. Record the new water level again. The difference between the two levels is the volume of the object.
Mass or weight*	Gram or kilogram	Balance or scale	Measure small quantities in grams or milligrams (one-thousandth of a gram or 0.001 grams). Use kilograms for larger quantities such as human body weight.

Variable	Unit	Measuring device	Notes
Temperature	Degrees Celsius (°C)	Thermometer, any of several different forms	Fever thermometers have a limited range and are not useful for most science projects. Ask your teacher about laboratory thermometers that can measure from below freezing (0°C) to above boiling (100°C).
Chemical content (for example, vitamin content of a food or concentration of a pollutant)	Usually, a mass per unit volume measure such as grams/serving or milligrams/liter	Often, specialized test kits or procedures that yield results in an appropriate mathematical form	Order test kits from science supplies houses, aquarium shops, swimming pool maintenance firms, and other sources.
Acid/base chemical characteristics	pH scale from 1 (very acidic) to 14 (very basic)	pH paper, liquid indicator, or laboratory pH meter	Make sure you understand what the pH scale means before you use it.

Light absorption, transmission	Camera f-stops; percent absorbency or transmission	Light meter; spectrophotometer	For simple light measurements, use a handheld light meter or an old-fashioned camera with a built-in light meter that requires manual setting of f-stops. A spectrophotometer is an excellent tool for measuring light absorption of liquids. It also permits chemical profiling by recording transmission/absorption at different wavelengths.
Sound	Decibel	Sound meter	The decibel scale is logarithmic. Understand it before you use it.

*On Earth, mass and weight are the same (at sea level). Mass is the amount of matter in something. Weight is the mass multiplied by the force of gravity. Since gravity's force varies only slightly with altitude while on Earth, a scale or balance that measures weight also measures mass. In space or on planets of different gravitational strengths from Earth's, mass and weight can differ greatly. Mass never changes, but weight varies widely depending on gravity's force.

Now determine which physical variables you want to measure and how you will measure them. Ask your teacher or an expert to help you learn the best way to take your measurements easily and accurately.

Measuring subjective variables

What if physical measurements don't suit your project? You don't want to measure distance, speed, noise, or acidity. You're more interested in people—what they think, how they feel, the things they do. Such variables can't be measured with rulers, light meters, or pH paper; but you *can* measure them.

Four tools for measuring subjective variables are

- observation
- existing records
- interview
- questionnaire

Each of these tools has its advantages and disadvantages:

Tool for data collection	Examples	Advantages	Disadvantages
Observation: Observe, count, and record for yourself.	Students' food choices in the cafeteria line. Weight of recyclable materials going out in household garbage. Colors of clothing worn in summer/winter.	If collected carefully and without prejudice, data are accurate and complete.	Observation requires both impartiality and anonymity. Believe strongly in what you are looking for and you'll certainly find it—even if it isn't there. Also, if your subjects know what you are doing, they may change their behavior.
Existing records: Some data already exist. No need to collect your own.	Crime statistics. Earned income (see volumes such as *Statistical Abstracts* or *Information Please Almanac* (most recent year). Incidence and frequency of diseases (from the Centers for Disease Control and *Prevention's Morbidity and Mortality Report*).	Existing data from reputable sources should be accurate and up-to-date.	Some data are confidential and you may be denied access. Examples include school attendance records; medical records; military, law enforcement, and certain court proceedings and records.

Tool for data collection	Examples	Advantages	Disadvantages
Interviews or questionnaires: Subjects self-report their attitudes, choices, or actions.	How much time eighth graders spend listening to the radio. Frequency of teens' church attendance. Parents' favorite book or movie. What fifth graders say they look for in a friend.	Interviews and questionnaires are an easy way to collect a great volume of data from a large number of people quickly and easily.	Caution: Interpret results with care. People may not lie intentionally, but they may say what they think they should or what they think you want to hear. It's only human to want to cast oneself in the best light.

If you plan to use observational techniques or existing records, your method of data collection may not be difficult to plan. With observation, you'll need to set up your procedure to avoid bias. For example, it wouldn't be fair to calculate the average number of cars passing a particular intersection from data collected only at 3 P.M. In the same way, it wouldn't be fair to

- assess children's yearly candy consumption from what they eat on November 1 (the day after Halloween)
- generalize about teens' choice of leisure clothing from observations made in a mall (What about all those who don't go to the mall?)
- conclude that people in your town throw away too much recyclable material based on what you found in one family's garbage

To make your test fair, you must be careful to collect the right kinds of data at the right times and places and in sufficient amounts. Most important, *don't jump to conclusions*. Your interpretations must be limited to what you *actually measured*, not what *you wish you could have measured*.

In most cases, existing data are reliable, appropriate, and bias-free if you obtain them from a source that is equally reliable, appropriate, and bias-free (see "How Useful Is It?" page 66). Make sure the data you use are up-to-date as well. Although five-year-old almanacs have a way of hanging around homes and libraries, they are useless for your science project. The same applies to statistics you obtain locally. Data on the number of robberies from a decade ago are great if you want to chart how crime statistics have changed year by year. If you want to assess what's going on now, however, you'll need numbers no more than a year old (or the latest available).

Surveys—including interviews and questionnaires—require the utmost care in both design and execution. When asking people questions—whether aloud or on paper—the answer you receive depends on the question you ask. If you load your questions, you bias your data. If you ask a fair question, you're on your way to a high-quality project.

> ALERT:
> Motto for interviews and questionnaires:
> Garbage In, Garbage Out

If the truth be told, conducting surveys is just plain fun. If you are planning to conduct interviews or hand out questionnaires, you may well feel excited. Finding out what other people think is interesting, something like tuning in on gossip or reading somebody's diary. Before you begin, however, make sure

- You have **permission**. *All* projects involving humans require permission.
- You have your **fair test** in order. If you ask different questions of different people, your data won't mean much. If you ask biased questions, you'll get biased answers. Either way, your science project goes right down the drain.

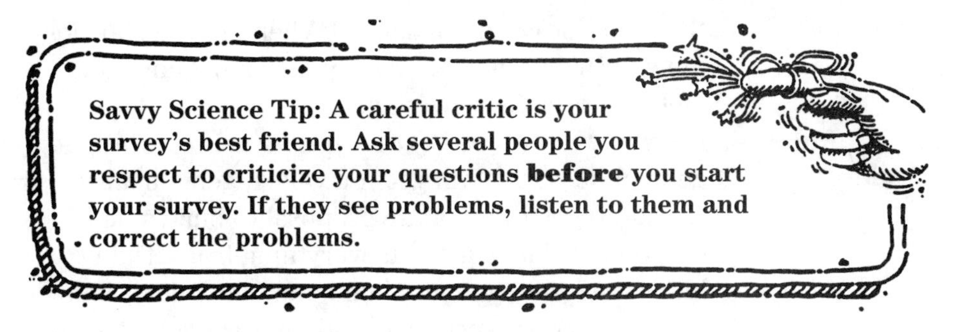

Savvy Science Tip: A careful critic is your survey's best friend. Ask several people you respect to criticize your questions before you start your survey. If they see problems, listen to them and correct the problems.

No magic potion makes good questions float and bad questions sink. However, testing your ideas against some rules may help sharpen your critical judgment.

Rule #1. *Ask for specific information.* "How do you spend your spare time?" may get you a wide variety of answers, but what will you do with them once you have them? Instead, target a specific subject such as "How many days a week do you go to the arcade?" "How much time do you spend at the arcade each week?" or "What are your three favorite games at the arcade?"

Rule #2. *Stay away from sensitive subjects.* Some matters are personal, private, and confidential; and people have a right to keep them that way. Don't ask for details of religion, health history, finances, or personal life. For example, your subjects may turn away from questions such as "Have you ever been tested for AIDS?" "Have you ever been arrested?" or "How much money do you earn in a week?" Not only will you alienate your subjects, you won't be able to trust your data. People may lie to avoid revealing personal secrets.

Rule #3. *Respect people's feelings.* Avoid questions that offend. A question you think is innocent enough can elicit a biased answer if your subject finds it sensitive. Young people eager to be adults may lie to "How old are you?" in certain situations. Middle-aged people may do the same. "How much do you weigh?" can threaten thick and thin alike. "When did you start losing your hair?" may not be a question balding men feel comfortable answering.

Rule #4. *Restrict the number of acceptable answers*. Vague, general questions will get you vague, general answers. "What do you think of the proposal to build a new courthouse?" may get you everything from "Didn't know they were going to build one," to "I don't know (or care)." More specific questions steer subjects toward answers you can use. For example, "The City Council is considering the construction of a new courthouse. If you were talking to your representative to the council about the proposal, would you speak for it or against it? Why?" You may still get some "I don't know" answers but not because your subjects haven't at least considered the situation.

Rule #5. *Set up categories for response*. Subjects may have an easier time answering your questions if you group possible responses into categories. You'll also find your data easier to interpret later when you make graphs and look for trends. For example, "How much time do you spend watching television?" may give you

answers that aren't very accurate and vary widely. A better question may be: "On an average school day, how much time do you spend watching television? (Check one):

A. less than one hour
B. one to two hours
C. two to three hours
D. more than three hours"

Rule #6. *Move questions from point A to point B one step at a time.* As you build your questionnaire or interview sheet, think of a logical progression of questions. If subjects answer "No" to an early question, the process ends there. However, if they know more or have more to share about the topic, they move on to more in-depth questions. For example, about the proposed new courthouse, you might ask first, "How many times in the last year have you been inside the old courthouse?" If the answer is "none," stop there. If "once or more," ask, "Did you see anything specific that you thought was wrong about the old courthouse?" If "yes," ask for a description. Then ask, "In your opinion, does this community need a new courthouse? Why?"

Rule #7. *Your survey's best friend is "Why?"* Never be content with a "yes" or "no" answer. Ask for reasons, either in categories you devise or in your subjects' own words. For example:

- Which candidate do you favor in the school elections? Why?
- Do you think people under age 16 should be able to sue their parents? Why or why not?
- Do you think the governor is doing a good job? Why or why not?
- Do you feel more safe or less safe in school this year than you did last year? Why?

Rule #8. *Avoid the "When did you stop beating your wife?" type of question.* Make sure your questions contain *no* assumptions about your subjects' actions, thoughts, or attitudes. For example:

BIASED:

Does the noise from the airport bother you more during the day or at night? (Your subject may not be bothered or may be bothered equally at all times.)

BETTER:

Do you notice noise from the airport? If so, does it bother you more at one time of the day than another?

BIASED:

In your opinion, why is our school's principal so unfair? (This assumes the principal is unfair by some objective measure and that your subjects agree with that conclusion.)

BETTER:

Do you think our principal is fair? Why or why not?

BIASED:

What do you think should be done to stop Acme Chemical from polluting the river? (This assumes that Acme Chemical is guilty of pollution and that your subject agrees with your judgment. It also assumes your subject agrees that some action should be taken—which isn't necessarily so.)

BETTER:

Do you believe that Acme Chemical is polluting the river? Why or why not? What, if anything, do you think Acme should do to improve the quality of water in the river?

Rule #9. *Order your questions from easy to difficult, non-threatening to controversial.* Ask the easy questions that don't touch nerves early. Once your subjects are hooked, then move on to the more difficult or potentially sensitive questions.

Rule #10. *Don't sound like a salesman or trickster.* If people think you are selling something or trying to set them up, you'll get little cooperation. Don't ask your subjects to reveal their phone numbers, home addresses, or names of family members. Don't ask respondents to promise some purchase or course of action. You'll spark suspicion if you ask people to disclose what computer equipment they own or how much money they earn.

Rule #11. *Make your identity and purpose clear.* Whether you approach people for interviews or distribute questionnaires, make clear—verbally, in writing, or both—exactly who you are and what you are doing. For example:

"My name is ____________________, and I am doing a project on ______________________________ for my ____________________ class at ____________________ School."

If interviewing, carry your school ID card. On questionnaires, give contact information for your school or teacher (with permission, of course) so skeptical subjects can verify your identity and your purpose.

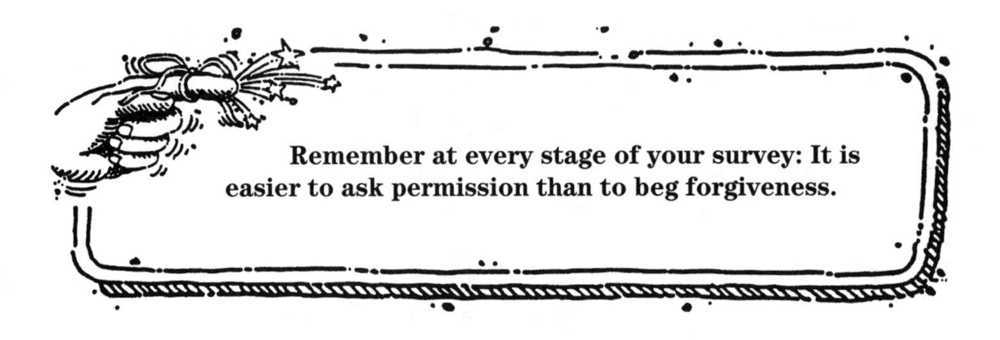

WAYS TO SAMPLE: HOW MANY ARE ENOUGH?

Each of the following projects has a serious flaw. Can you spot it?

- Marissa plants two seeds in separate pots. She puts one in the light and the other in the dark. She measures growth every day for two weeks.
- Ben wants to know if boys and girls like the same colors. He asks his brother and sister about their favorite colors. His data show that boys like red and girls like blue.
- Sharyl takes two pictures of her mother. She uses the same kind of film in two different cameras. One is a cheap model; the other, expensive. When the pictures are developed, she asks her brother if he can tell which picture is which. He says he can't. Sharyl concludes that the cheap camera is just as good as the expensive one.

Beware the deadly one-sies. One of anything is never enough.

These students have fallen into the trap of the "deadly one-sies." Each wants to **draw conclusions**—make general statements of truth—but none can. Why? In each case, the student measured only *one* of something.

- Marissa placed *one* plant into the light and *one* plant into the dark. No matter what happens with the plants, what can Marissa conclude? Marissa can only show that she didn't study enough plants to draw any generalizable conclusion.
- Ben asked *one* male respondent and *one* female respondent. From a sample of one, Ben cannot fairly conclude anything about general differences between all boys and girls.
- Sharyl took *one* picture using each camera. *One* person judged the quality. Sharyl has too few pictures and too few judges to compare the cameras fairly.

Why are the one-sies so deadly?

Chance.

Everything varies in random, unpredictable ways. No two cases of anything are exactly alike, whether you're studying circuits, sow bugs, or senior citizens. Chance or random variation occurs for no particular reason. It just happens.

Variation is the nature of nature.

Convince yourself that's true with a simple experiment. Get a pen and paper, and quickly sign your name ten times. Study the signatures. No two are exactly alike. You didn't try to make them different. You thought they were all the same; but they vary subtly, one to the next, simply by chance.

You couldn't pick out one of those signatures and say it (and only it) was yours. Your true signature is closer to an average taken across 10, 100, 1,000, 10,000, or a lifetime of signatures.

Samples and populations

Because one is never enough, an investigator must *sample* from a *population*.

Sample: A small amount or number taken from a larger quantity or number.

Population: The total of all things in a category.

Some examples:

POPULATION:

All seventh graders at Krieger Middle School

SAMPLE:

Fifty seventh graders selected randomly from all seventh graders at Krieger Middle School

POPULATION:

All hot fudge sundaes in Meterville

SAMPLE:

Two hot fudge sundaes purchased at each of six randomly selected sites in Meterville

POPULATION:

All parallel circuits in existence or ever made

SAMPLE:

Six parallel circuits constructed for test purposes

Note that you must define your population as part of your question. If you want to be able to generalize about the population of earthworms inhabiting your backyard, then select from only your backyard. If you want to be able to say something about all the earthworms in town, then sample widely from several different sites.

Norm's Crickets: A Case Study

Norm learned about sampling the hard way. He built a shoebox habitat with a light side and a dark side. He put a cricket into the box and watched to see toward which side the little animal would move. He recorded what happened, wrote his report, and made a pretty poster.

When his teacher asked, "What can you say about whether crickets move toward light or dark environments?" Norm gulped. He realized he couldn't say anything. He knew what one cricket did one time, but he had no idea what all the crickets in the world do most of the time. Norm fell prey to the deadly one-sies. One cricket. One trial. He used too small a sample to make any generalizations.

Norm needed to sample. He needed to work with as large a number of crickets as possible and to see how they respond *on the average.* Because he couldn't possibly study all the crickets in the world (the *population),* he needed a sample—a smaller number of crickets that might *fairly* represent the population.

What is a fair sample?

A fair sample is an unbiased, *random* selection from the population. In these pairs, which sample (indicated by the Xs) is fair and unbiased?

Potholes in a road

People in a community

In the first drawing, the selection of the largest potholes is unfair.

In the second, the selection of only females is unfair—unless the test population is limited only to females.

Consider these examples:

QUESTION: How does the volume of leguminous seeds change during germination?

Poor sample: One bean seed and one pea seed
Better sample: Twenty seeds each: bean, pea, lentil, alfalfa, and soybean. (Note that the question is specific and limits the study to only leguminous seeds. To attempt to generalize about seeds of every kind would be too much to take on in one project.)

QUESTION: How many different kinds of organisms live in the soil around Stow's Pond?

Poor sample: One 100-gram sample taken one meter from the pond's edge on the north side.
Better sample: Eight 100-gram samples taken at *randomly selected* sites on all sides of the pond and at distances varying from one to ten meters from the pond's edge. (Note: The display and written report should include a map of sampling sites.)

QUESTION: How does ultraviolet radiation affect the growth of mold on white bread?

Poor sample: One slice of white bread irradiated for three minutes; one slice of nonirradiated bread (as a control)
Better sample: Five groups of white bread slices, each containing three slices; Groups irradiated for 0 (control), 3, 6, 9, and 12 minutes. (Note: By using identical numbers in each group, the project might be expanded to study other kinds of bread such as rye, pumpernickel, or sourdough.)

What does random mean?

Whenever sampling is mentioned, the word *random* is likely to pop up. *Random* means *by chance*. You must not *choose* the members of your sample for any particular reason. In truly fair samples, the individuals in the sample have been included merely by chance. For example

Random	Not random
Boys whose names are drawn from a hat containing the names of all boys in the entire student body	Names drawn from a hat containing names of members of the football team (unless the football team is the population)
Strangers approached on a street corner	Your family or friends
Every third pea taken from 100 pods	The biggest peas taken from 100 pods
Twenty-milliliter samples of three shampoos randomly selected by putting all known brand names into a hat and drawing out three	Twenty-milliliter samples of your favorite three shampoos

How can you be sure your samples are random? Find a way to remove any bias or prejudice from the selection process. Don't choose for a reason. Let luck make the selection for you. For example, when measuring objective variables in a land use study, you might

- Drop a pencil onto a map to select a site.
- Take multiple samples at different times, sites, or depths (as in water), choosing sampling conditions by the toss of a coin.
- Tie a string between two stakes driven into the ground at points *a* and *b*. Sample at sites a preset distance along the string, for example every 10 meters. (Ecologists use this technique frequently. It is called a line transect. For small study sites, use balsa wood or meter sticks to make a square that measures a meter on each side. Toss the square randomly into the site, and study everything you find within the meter-square area.)

For subjective variables:

- Write the names of individuals on small pieces of paper. Put all the papers into a container. Draw out a small number. No peeking and no returning choices you don't like!
- Assign numbers to individuals. Draw numbers without peeking.
- If interviewing people on the street, toss a coin. If heads, ask for an interview. If tails, let them pass by.
- Select every one-hundredth number in the telephone book. (This does not quite create a random sample, as it excludes people without phones or with unlisted numbers.)

How many are enough?

No magic formula will reveal how large your sample size needs to be. An adequate size sample depends on

- the nature of your project;
- the size of your population;
- how many subjects you can reasonably handle; and
- how far you want to go in generalizing your findings.

If the population of students in your school is 2,000, then ten is hardly a large enough sample to tell you anything about students' attitudes in general. Fifty or 100 may reveal more, but anything over 200 probably isn't necessary. When the sample gets very large, the **law of diminishing returns** starts to set in: You expend more and more effort and get less and less new information. You'd likely find that your first 100 respondents tell you nearly everything you would hear from the remaining 1,900. Still, the only way to describe the population with complete accuracy is to survey all 2,000 members.

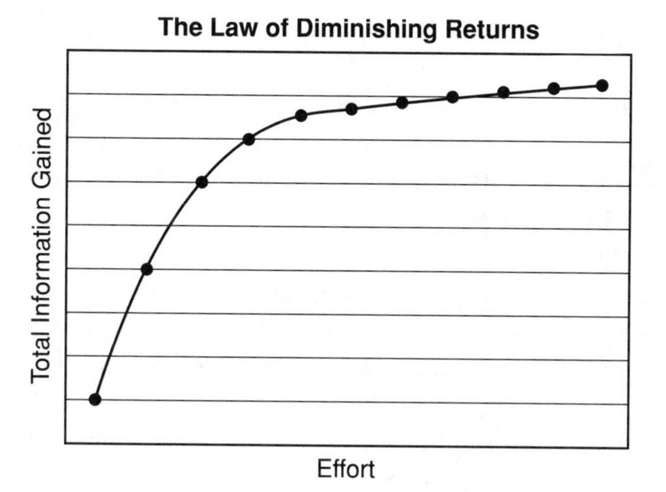

Practical matters limit sample size, too. You might be able to plant 100 radish seeds easily, but you can't manage 100 azalea bushes unless you work in a greenhouse. For a study of rabbits or cats or pet iguanas, a half dozen might be a reasonable number. If you work with bacteria, however, expect to count colonies by the thousands.

How far you want to go with your conclusions may tell you how big to make your sample. Don't expect to generalize to all gardeners in your state by interviewing only those in your neighborhood. You can't tell what all teenagers think if the only ones you interview go to your school. You can't determine the ideal growth conditions for all plants from a study of bean seedlings alone. Limit your topic, and you'll see ways to limit your sample to a manageable size.

General rules for determining sample size

1. The *larger the sample, the more nearly accurate the measurement* of the dependent variable.
2. The *more the members of a population vary, the larger the sample must be.* If you want to study all eighth graders in your school, you'll need a larger sample than if you are interested in generalizing only to eighth-grade girls who run track.

3. The sample size has to be *big enough to allow the grouping or categorization* of subjects and findings. For example, a student who plans to compare two growing conditions for radish seeds may be content with a sample of 60 seeds, 30 in each group. To compare six growing conditions, however, you will need 180 seeds to maintain six groups of 30 each.
4. Use the rule of threes. *When in doubt, go for multiples of three.* Don't conduct one trial—conduct three, six, or nine. For sample sizes, add a zero and consider groups of 30, 60, or 90.
5. If your data seem too spread out, increase your sample size. In general, *larger samples reduce variation and reveal trends better* than small ones. Compare these examples:

Intended Vote on Amendment 12: Ten People Interviewed

	Vote			
Voter's Sex	**Yes**	**No**	**Undecided**	**Total**
Male	2	0	3	5
Female	1	3	1	5
Both Sexes	3	3	4	10

A graph of these data shows no clear trends. The sample is too small to predict whether men and women will vote differently or whether the amendment will pass.

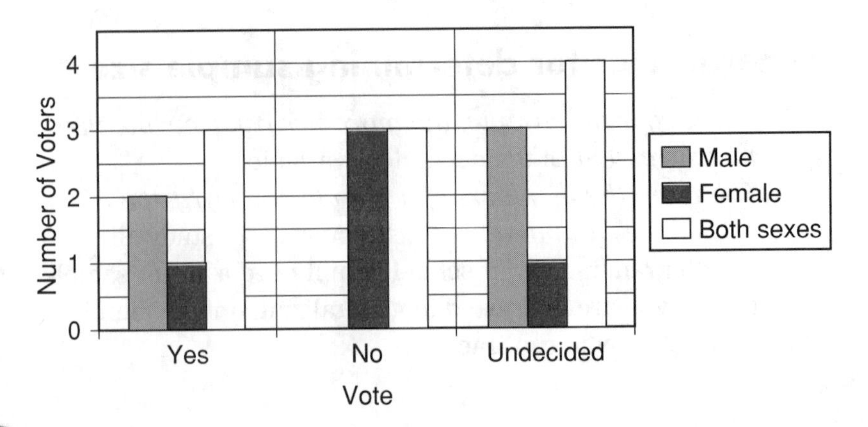

With a larger sample, however, trends begin to appear.

Intended Vote on Amendment 12: 100 People Interviewed

	Vote			
Voter's Sex	**Yes**	**No**	**Undecided**	**Total**
Male	26	18	6	50
Female	15	30	5	50
Both Sexes	41	48	11	100

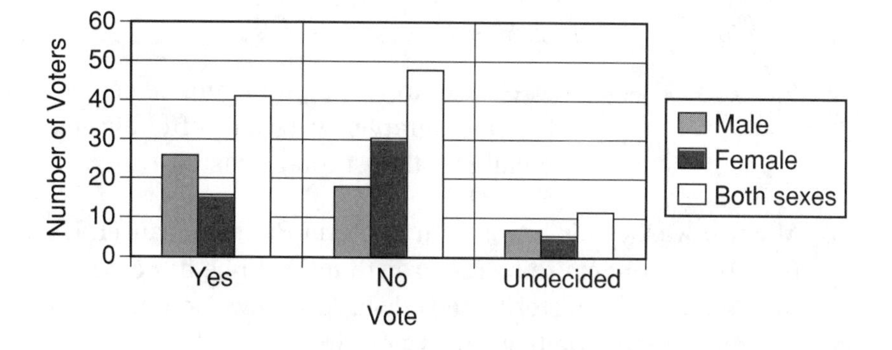

The larger sample shows that

- men favor the amendment more than women do;
- if men and women vote in equal numbers, the amendment will probably fail although not with a majority of the vote; and
- the undecided voters could swing the vote either way.

BRAIN TICKLERS
Set # 12

Fix each of the following projects by correcting sampling procedures, variables, or both.

1. Arthur microwaves five radish seeds at 100 percent power for five minutes. He plants them. None sprout. He concludes that microwaves kill seeds.
2. Teri puts an iron nail into lemon juice and takes photographs every day. After a week, the nail has almost totally rusted.
3. Michael washes his hands with Bacteria-Bubble cleanser. He touches his fingers to sterile growth medium he has carefully poured into a sterile petri dish. After a week, only a few colonies of bacteria have grown on the agar.
4. Aisha stains two pieces of white cotton fabric with grape juice. She washes one in Sudsy-Soap, the other in Ditsy-Detergent. The one washed in Sudsy-Soap comes out whiter. She decides she can recommend Sudsy-Soap as the better stain fighter.
5. Perry picks up cones from beneath six spruce trees and six pine trees. He gets an average of 23 from each spruce. He gets an average of 17 from each pine. He reports that spruce trees drop more cones than pine trees.
6. Nikita plants a kidney bean upside down and a pinto bean right side up. The kidney bean comes up first. She decides that beans sprout faster if they are planted upside down.

7. Steve puts objects onto trays and shows them to 15 boys and 15 girls, each for one minute. The boys see trays of 10 red objects. The girls see trays of 10 different cosmetic products. After removing the trays from sight, Steve gives each person one minute to recall as many items as possible.
8. Marissa puts 200-speed Kaibab film into a Pentarc camera. She puts 400-speed Fujiyama film into a Canyon camera. After the pictures are developed, everyone agrees the Pentarc pictures are sharper.
9. Dylan builds a model arch bridge using 100 paper straws and a model beam triangle bridge using 100 plastic straws. The beam triangle bridge holds more weight before collapsing.
10. Megan asks seventh graders which class they like better: math or science. She asks eighth graders if they prefer math or social studies. She finds that seventh graders like math and eighth graders don't.

(Answers are on page 133.)

Finding research subjects

If you are planning to program a computer to solve equations or build a model of an earthen dam for your science project, then you needn't worry about populations, sampling, or finding research subjects. If, however, your project requires

- plants,
- animals, or
- people,

then you must think about how and where you will get your participants. If you're growing your own plants for your project, all you'll need to do is plant plenty of seeds early enough to ensure an adequate sample size.

If, however, you want to study plants in a natural or human-made environment, such as a forest or a park, you'll need permission from the owner or caretaker of the land. Never pick flowers or dig up plants without permission. Never go onto private or publicly owned land unless you are sure it is OK to do so.

The same cautions hold for animals. It is probably OK to collect some earthworms from your backyard for your experiment so long as you treat them gently and return them to their natural environment quickly. You cannot, however, do that on other people's land without permission, nor can you go after larger animals such as cats or dogs. If in doubt, ask your teacher or project advisor.

Finally, your project may require human subjects. Projects that investigate people always require permission from a review committee. The committee will want to know

1. *How you plan to recruit your participants.* Suppose you want to study how babies react to different kinds of music—classical, country and western, or heavy metal. Finding the music is no problem, but where can you find babies? Start with people you know. Ask friends, relatives, or teachers to help. Maybe parents in your neighborhood could get involved, or you might contact a local day care center or nursery.
2. *What you plan to ask or do with your subjects.* In our example of babies and music, researchers often measure how long babies' eyes stay focused on an object as a indication of attention. With surveys, you may need to specify what questions you will ask or what task you will present. Examples of the latter include memorizing nonsense syllables, recalling objects seen on a tray, or repeating numbers in a series.

3. *How you intend to use your results.* In all cases you must protect the confidentiality of your subjects. You must not use their names in your project. You must not reveal individual opinions, responses, or performance. Be ready to show your review committee what kinds of tables and graphs you expect to create from your data. (See Chapter Four.)

BRAIN TICKLERS—THE ANSWERS

Set # 10, page 92

1. I = amount of glycerin; D = time before bubble burst; C = temperature, wind, size of bubble blown, type of bubble blowing equipment, place tested, stopwatch used, and many more

2. I = temperature of water; D = amount of salt dissolved; C = amount of water, containers, thermometer used, heat source, testing conditions, measuring tools for salt and water, and many more

3. I = two (or more) different brands of cola in regular and diet form; D = answers from taste-testing volunteers; C = size of samples, brands of cola and their source, instructions to subjects, testing conditions and apparatus, and many more

4. I = kinds of plants; D = whether they grow stems; C = pots, soil, water, light, time and method of cutting, times of observation and data recording, and many more

5. I = circuit arrangement, series and parallel; D = time before bulbs burn out; C = brand and size of bulbs, kinds and lengths of wires, mountings, switches, test conditions, times and conditions of observation and data recording, and many more

6. I = corn syrup and glycerin; D = time to fall a measured distance; C = size, shape, material and weight of bearing; brand and source of corn syrup and glycerin; test chamber height, material; volume of corn syrup and glycerin samples; distance the bearings fall; timing device; timing conditions; and many more

7. I = variety of disinfectants and odor-control devices; D = numbers of bacteria grown from source; C = type of shoes tested, test conditions, methods of growing and counting bacteria, time, temperature, and many others

8. I = age of person interviewed; D = reported number of hours slept per night; C = interview questions and techniques

9. I = phases of the moon; D = barometric pressure; C = ??? The investigator cannot control such variables as season of

the year or weather conditions. Also, since data may come from the calendar and the weather service, the investigator may not be able to control data collection procedures. The only solution is to collect data over a very long period of time and use mathematical techniques (statistics) to look for relationships.

10. I = shower water and tap water (or whatever is normally used to water the lawn); D = some measure of the health of the test grass such as color, thickness, height, growth rate, or something else; C = the kind of grass tested, the source of the shower water, the testing procedures and measuring tools, test conditions including temperature, growth time, and exposure to light, and many more.

Set # 11, page 95

1. Stacey needs to test samples of all three kinds of milk—including 2 percent—and to specify precisely what it is about milk that she wants to study. She must also find a way to *quantify* her results (use numbers) to make meaningful comparisons. If taste is what she wants to measure, then taste tests and ratings (on, perhaps, a scale of 1 to 10) by blindfolded volunteers might help answer her question. However, if shelf life or spoilage rate really interests her, then she might record the number of days that pass before the first signs of spoilage appear. She might also learn how to grow microorganisms from the milk on growth media and count the numbers of bacterial colonies present as the days go by—being sure that all her samples are equally fresh when she begins.

2. All Hobart is showing is the amount of rain that fell in April. To test his hypothesis, he needs rainfall measurements from the other 11 months of the year. He would be better off, also, with data for more than one year. In addition to taking his own rainfall measurements, he should call the weather bureau for numbers from years past.

3. If Kelly defines "best" as slow running, she needs to time many different brands repeatedly (averaging results from multiple trials) in order to compare them. She should recognize that other measures might well be used to define quality—perhaps taste tests, sugar content, or shelf life.

4. Coffman needs a control and some comparisons. How tall do the same kind of tomatoes grow without Vital-Grow? How tall do they grow with other kinds of fertilizers? Is height the best measure of success with tomato plants?

5. How many slips did teachers write in other months? If the numbers are highest in the winter months, does that necessarily mean that cold weather is the cause? Some careful questioning of students, teachers, and the principal might reveal other important factors such as special events in school, exam stress, or parental pressure.

6. Monty needs samples from other ponds that are less acidic. He also needs some measurement of the numbers and varieties of living things in his samples in order to make comparisons. Without more data, he can't assume that pH differences result from acid rain. Natural factors or other sources of pollution may be at work.

7. Paige needs to develop some specific measure of shampoo quality such as how users rate their hair's shine. She also needs to include shampoos other than Sunset for comparison. Paige should involve more than three subjects in her experiment. Finally, she should use subjects who aren't her friends (and not so likely to tell her what she wants to hear).

8. Dion can't generalize to all trees. He extracted pigments only from maple leaves. He also cannot assume that cold weather is the reason he found no chlorophyll.

9. Happiness is a condition a human being can report verbally. Chameleons lack that ability. Samantha should report only what she can actually see or measure—for example, an

animal's color immediately after eating or at a certain air temperature.

10. Martin has not tested color preferences. He has compared food choices, which might not be the same on different days or with different tortoises. To compare color choices, offer the same food dyed different colors, and test as many tortoises as possible over a long time period.

Set # 12, page 126

1. Use groups of perhaps 20 or 30 seeds each. Microwave at different power levels for different periods of time. Arthur cannot draw conclusions about any other kind of seeds—only radishes—unless he actually tests other kinds.

2. One nail in one container of lemon juice is not enough. Several setups (perhaps three or four) should show whether all iron nails rust under this condition. Teri needs a control—in this case, nails in identical containers but containing no liquid. Nails exposed to water for the same period of time make a good basis for comparison (control) also. Teri cannot generalize to all nails, only those made of iron.

3. One trial is not enough. Set up five or six dishes in each of three groups: (1) plates untouched by Michael's hands; (2) plates touched by Michael's fingers before he washed; and (3) plates touched after washing. The control group (1) shows if bacteria grow on untouched plates. The experimental groups (2) and (3) show if the cleanser actually makes any difference.

4. We must know for sure that Aisha stained the same kinds of fabric with the same juice for the same length of time. The sizes of her fabric samples and the washing procedures must be identical, too. Assuming Aisha has controlled all those variables, she needs only to increase her sample size—washing perhaps five or six samples in each

detergent—and to turn her judgment of "whiteness" into a number. She cannot, however, even when all these steps are taken, conclude that one detergent removes all stains better than the other. If she tests only grape juice, then grape juice is the only stain she can compare. For a better project, she might decide to test grease, ketchup, and chocolate stains, too.

5. Too many variables may have been left uncontrolled here. Are the trees all the same size and age? If not, do both groups vary about equally in size and age? Did Perry collect all the cones at the same time or over the same time period? Collecting beneath one kind of tree in September and the other in April is not a fair test.

6. Nikita introduced an unwanted variable: the kind of bean. If she wants to compare right side up with upside down, she must use the same kind of bean in both setups. She also fell prey to the deadly one-sies. Beans are cheap, easy to get, and take up little space. Plant 20 or 30 of each, record the time until germination of each. Average the results.

7. If Steve wants to see whether boys or girls remember objects better, he should show both genders the same trays. If he wants to see if red objects are more memorable than cosmetics, he should show each tray to both boys and girls.

8. Marissa actually has three independent variables in her experiment, but she is drawing conclusions as if she had only one. Her three variables are (1) speed of film: 200 versus 400; (2) brand of film: Kaibab versus Fujiyama; and (3) make of camera: Pentarc versus Canyon. If she wants to draw conclusions about the make of camera, she should use the same brand and speed of film in both cameras. Marissa can test more than one independent variable at a time, as long as she includes all possible combinations in the experimental design. In this case, that would mean testing four types of film (Kaibab 200 and 400; Fujiyama 200 and 400) in each make of camera.

9. All experimental variables must be kept constant except the desired independent variable. If Dylan wants to compare arch and triangle bridges, then both should be built of 100 of the same kind of straw—either plastic or paper. If, on the other hand, Dylan wants to compare the strength of plastic and paper straws, he should build identical bridges from the two different materials.

10. Both seventh and eighth graders need to be asked the same question. Since the nature of math, science, and social studies classes varies between grades, the results may not mean much. A more meaningful study may be students' performance in different kinds of classes or the career fields they say interest them.

CHAPTER FOUR
Collect and Analyze Data
Should have collected enough data by now?!

Perhaps you feel nearly ready to begin collecting data. Maybe you have even started and found a few problems. Now's the time to get your experimental procedure pinned down and to smooth out any rough spots in your project design.

WRITING A PROCEDURE

Your **procedure** includes your materials and all the steps that you follow in setting up your experiment and collecting your data. You should write your step-by-step procedure in your journal and follow it to the letter.

If you change your procedure, you will change your project—you hope for the better. Later, when you prepare your final report, you'll want to tell the world how you improved your procedure as you went along.

What's included in a procedure?

Write your procedure as you would write a recipe:

- List all materials, equipment, and supplies needed. Leave out nothing.
- Describe each step from beginning to end—in order—again leaving out nothing.

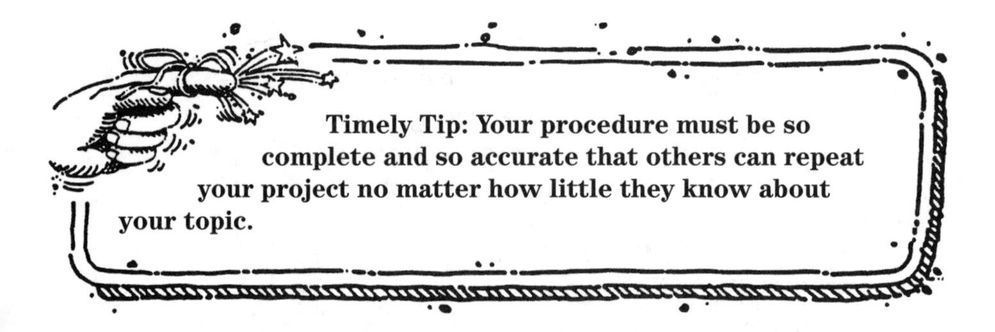

To understand the importance of a complete and stepwise procedure, imagine yourself as an alien who has just landed on Earth from the Planet Ignoramia. You know Earth language, but you know nothing of the habits and behaviors of humans. You notice an Earthling eating something.

"What's that?" you ask.

"A peanut butter and jelly sandwich, you Ignoramus," replies the Earthling with a scowl.

"How do you make a peanut butter and jelly sandwich?" you ask, not offended at the mention of your planet.

Which of the following responses would help you most? Why?

RECIPE 1:

Spread peanut butter and jelly on two slices of bread. Eat. Enjoy. Answer stupid questions from alien invaders.

RECIPE 2:

Materials:
Loaf of fresh bread (from the bread box)
Knife (from the cutlery drawer)
Jar of peanut butter (from the cupboard)
Jar of jelly (from the refrigerator), any flavor, but grape is very good
A plate (from the cupboard)
A paper towel (from the counter)

1. Assemble all materials. Open the bag the bread is stored in.
   Take out two slices.
   Place the slices on the plate side-by-side.

2. Remove the lid from the peanut butter jar by turning the lid counterclockwise. While grasping the knife by its handle, dip its flat surface into the peanut butter, scooping some of the peanut butter onto the knife's flat surface.
3. Spread the peanut butter onto the surface of one of the bread slices. Get more peanut butter from the jar, and continue spreading until one slice of bread is totally covered. Place the slice onto the plate.
4. Repeat steps 2 and 3 with the jelly and the other slice of bread.
5. Lift the jelly-covered slice and turn it over, lowering it to rest atop the peanut butter.

6. Clean the knife and countertop with the paper towel. Replace the lids on the jars. Return the peanut butter to the cupboard and the jelly to the refrigerator. Return the bread to the bread box.
7. Eat the sandwich and enjoy it. What should you do with the plate? See the recipe for "Washing and Drying Dishes."

The second example is better because it

- lists every item needed to achieve the objective;
- gives step-by-step instructions;
- leaves nothing out;
- puts everything in the proper order; and
- would communicate with anyone who could read, not just someone who already knew a lot about peanut butter and jelly sandwiches.

Caution—Major Mistake Territory!

Students often make the mistake of leaving steps out of their procedures.

Their reason? "Well, anybody would know that!"

Not so! To be sure that anyone could repeat your procedure, pretend you are writing for an alien from the Planet Ignoramia.

Test your procedure

A good way to test the completeness and accuracy of your procedure is to try following it step-by-step, doing nothing except what the procedure says. If you have left gaps, you'll find yourself missing materials or wondering how to get from one step to the next.

Another good test is to trade procedures with friends. You should be able to follow theirs with ease, and they should be able to follow yours. If not, fill in the gaps and try again.

Hint: A photograph or a few sketches can go a long way toward making your procedure clear and easily understood by anyone.

BRAIN TICKLERS
Set # 13

Write a procedure for planting a seed in a pot.

(Compare yours with the sample on page 201.)

Getting organized

What if you went into the kitchen to make a peanut butter and jelly sandwich, only to find the bread box empty, the jelly moldy, and the knife missing from the drawer? Your procedure will work only if you get your materials easily at hand, clean, and in good repair.

- Gather all your materials.
- Make sure all your tools are in good working order.
- Learn to use all your equipment properly, including measuring devices.
- Ask for expert help in learning how to handle safely any hazardous material your experiment may require.

Caution—Major Mistake Territory!

Never try to fix equipment you think might be broken without help from an expert. The machine may be fine. You might be using it incorrectly. If the equipment is truly malfunctioning, you might unintentionally make the problem worse.

DATA TABLES

Part of your advance preparation involves deciding ahead of time how you will record your data. Make blank **tables** for recording your results, and then fill them in as you go. Deciding in advance on the form your tables will take may even help you see a better way to conduct your experiment.

What's a table?

No, not that kind of table!

For your science project, **a table is a series of rows and columns that show how two or more sets of numbers are related.** For example:

	Kind of Cereal			
	Corn Flakes	Corn Bran	Bran Flakes	Raisin Bran
Serving Size (weight in grams)*	30	27	30	59

*Data from the nutrition analysis on the package

Tables are always constructed in a standard way. The variable that you decide to study (in a nonexperimental project) or change (in an experimental one) goes across the top or side. The outcome measure (dependent variable) goes on either the top or side—whatever is left. Your actual measurements of the dependent variable go in the boxes. How you arrange the table is up to you, as this alternative shows:

Kind of Cereal	Serving Size (weight in grams)*
Corn Flakes	30
Corn Bran	27
Bran Flakes	30
Raisin Bran	59

*Data from the nutritional analysis on the package

Both of these tables show that the variable chosen for study was the kind of cereal. The outcome measure was the serving size (from the label, as recommended by the manufacturer). The unit of measurement for the outcome is part of the label (in this case grams, a measure of weight).

Tables may include more than one outcome variable, for example:

	Kind of Cereal			
	Corn Flakes	Corn Bran	Bran Flakes	Raisin Bran
Serving Size (weight in grams)*	30	27	30	59
Serving Size (volume in cups)*	$1\frac{1}{3}$	$\frac{3}{4}$	1	1
Calories Per Serving*	110	90	110	190

*Data from the nutritional analysis on the package

Calculations that reveal important ideas derived from data may be included in tables too.

	Kind of Cereal			
	Corn Flakes	Corn Bran	Bran Flakes	Raisin Bran
Serving Size (weight in grams)*	30	27	30	59
Calories per Serving* (volume in cups)*	110	90	110	190
Calories per Gram of Cereal†	3.7	3.3	3.7	3.2

*Data from the nutritional analysis on the package

†Calculated by the experimenter

In this example, notice how useful the calculation might be to people watching their weight. Even though raisin bran appears from the Calorie count alone to be the worst choice for losing weight, it's actually the best—*if* the individual weighs out a small portion.

For people more inclined to dump some cereal into a bowl, a calculation of Calories per unit volume might help more:

	Kind of Cereal			
	Corn Flakes	**Corn Bran**	**Bran Flakes**	**Raisin Bran**
Serving Size (volume in cups)*	$1\frac{1}{3}$	$\frac{3}{4}$	1	1
Calories Per Serving*	110	90	110	190
Calories Per Cup	82	120	110	190

*Data from the nutritional analysis on the package
† Calculated by the experimenter

From this point of view, corn flakes look like the dieter's best friend.

Do design, computer, and construction projects need tables? You bet! Every time you change a component or variable in your design, you'll want to record the outcome. What better way to do so than with a table?

Building tables to match your question

Your data tables must match your project design. Look again at your dependent and independent variables. How will they fit into a table? Study these examples from Brain Ticklers Set # 10 in Chapter Three.

EXAMPLE:

How does the amount of glycerin added to the bubble mix affect the length of time a bubble will last? I = amount of glycerin; D = time before the bubble bursts

Amount of Glycerin (milliliters per liter of bubble soap)	Time Before Bubble Bursts (in seconds)			Average of Three Trials
	Trial 1	Trial 2	Trial 3	
10				
20				
50				

Notice in the example:

- The table clearly states the units of measurement (milliliters per liter and seconds) for both variables.
- This experimenter has defeated the deadly one-sies. The table includes space for three trials.
- The last column allows for a calculation—in this case, the average of the three trials.

Some experiments need more than one calculation.

EXAMPLE:

Can volunteers tell the difference between diet and regular colas? I = two (or more) different brands of cola in regular and diet form; D = answers from taste-testing volunteers

Taste Tester	Cola-la Regular Correct Answer?	Cola-la Diet Correct Answer?	Coco-oh Regular Correct Answer?	Coco-oh Diet Correct Answer?	Total Number of Correct Answers
Zachary	yes	yes	yes	no	3
Kyla	no	no	yes	yes	2
Katelyn	yes	no	no	no	1
. . . and 36 more . . .	. . .	. . .	. . .	. . .	. . .
Total Correct Guesses	15	24	19	20	78
Percent Correct (number of correct responses divided by the number of respondents) $n = 39$	38%	62%	49%	51%	Average number correct per respondent: 2

In this experiment, the investigator calculates totals after recording answers from individuals. In addition,

- The table reveals that 39 subjects participated in the taste test. (By convention, the letter n stands for the total number studied in any experiment.)
- The totals of the **rows** show individual scores. The average tells the experimenter that, in general, the subjects guessed two sodas right and two wrong.
- The totals of the **columns** show how many times (out of 39) the taste tester correctly guessed the cola. The numbers don't mean much until converted to percents for comparison (see "Number Crunching I: For Beginners" starting on page 153).

BRAIN TICKLERS
Set # 14

Before making data tables for your project, sharpen your skills with the examples that follow.

Set A. Write questions that could be answered with data from each of the following five tables:

1.

Material	Light Meter Reading
Nylon stocking	
Fishing net	
Lace curtain	
Brown paper	
Plastic wrap	

2.

Weight of Ball Bearing (grams)	Height of Drop (centimeters)	Diameter of Crater (centimeters)
30	20	
30	40	
60	20	
60	40	

3.

Date	Time	Relative Humidity	Barometric Pressure	Number of Traffic Tickets in Meterville

4. Number of Students Passing Color Blindness Test

Grade	Boys (of 30 tested)	Girls (of 30 tested)	Total (of 60 tested)
Seventh			
Eighth			
Ninth			

5. Aluminum Content (in milligrams per liter)

Food	Storage Method	Day 0	Day 5
Canned tomatoes	In can		
Canned tomatoes	In plastic container left open		
Canned tomatoes	In closed plastic container		
Fresh tomatoes	In aluminum pan		
. . . and so on . . .	. . . and so on . . .		

Set B. Draw a table for recording data from each of the following experiments:

6. Brittany wanted to find out which flower-drying method removed the most water from flowers over a two-week period. She weighed zinnias and marigolds. She dried them in salt, borax, and silica gel.

7. Jared collected seashells from two different beaches and categorized them by color, weight, and shape.

8. Quincy wanted to compare the fat and water content of four kinds of ground beef. He weighed 100-gram samples of ground beef labeled *ground top round*, *ground bottom round*, *ground chuck*, and *ground sirloin*. He microwaved each sample on a paper plate for five minutes then weighed them again. (Hint: Don't forget the unit of weight.)

9. Courtney knew that dropping a slice of potato into hydrogen peroxide produces bubbles. She cut potato cubes of different sizes and dropped them into jars containing peroxide solutions of different strengths (1, 2, and 3 percent). She recorded the time that passed before the solutions stopped bubbling.

10. Russ hypothesized that the more time people spend watching television, the less time they spend reading. (Suggestion: Make a basic data table for recording data. Then make several others Russ might use to compare the watching/reading habits of groups. For example, Russ could compare teens and adults or boys and girls.)

(Answers are on page 202.)

NUMBER CRUNCHING I: FOR BEGINNERS

When trying to get meaning from your data, you may find that raw numbers tell you little, while some simple **calculations** reveal all.

Take Connor's project as an example. Connor wanted to find out whether plastic bags or paper bags decompose better when buried in soil. He did all the right things. He used several samples of each kind of material and controlled all other variables, including the length of time of burial (three months) and the soil in which he buried the samples. He weighed his samples before and after burial. Realizing that the amount of water the materials took up while buried in the soil would affect his outcomes, Connor made sure his samples were dry and clean before weighing. His first data table looked like this:

Material	Dry Weight Before Burying (in milligrams)	Dry Weight After Three Months (in milligrams)
Sample plastic—1	134	122
Sample plastic—2	143	117
Sample plastic—3	129	119
Sample plastic—4	137	125
Sample plastic—5	147	136
Sample paper—1	339	223
Sample paper—2	347	221
Sample paper—3	376	211
Sample paper—4	388	233
Sample paper—5	355	249

Connor wondered what to do next. By just looking at the numbers, he could tell that the paper bags lost more weight than the plastic. But how much more? Was the difference big enough to mean anything? How could he compare the changes in some precise way?

Addition

Connor started with simple totals. He added together the weight of all the plastic samples, both before and after burying. He did the same with his paper samples. He realized that totals across the rows would be meaningless. He also realized that totaling all plastic and all paper—before and after burying—would do him no good.

Material	Dry Weight Before Burying (in milligrams)	Dry Weight After Three Months (in milligrams)
Sample plastic—1	134	122
Sample plastic—2	143	117
Sample plastic—3	129	119
Sample plastic—4	137	125
Sample plastic—5	147	136
Total Weight: Plastic	690	619
Sample paper—1	339	223
Sample paper—2	347	221
Sample paper—3	376	211
Sample paper—4	388	233
Sample paper—5	355	249
Total Weight: Paper	1,805	1,137

Subtraction

The totals made the differences look larger, but they were still hard to compare. The change was bigger for the paper bags, but their beginning weight was bigger, too. Connor asked himself, "How much did the paper and plastic change over three months?" He knew that subtraction reveals a difference or a change. He subtracted and added a column to his table:

Material	Dry Weight Before Burying (in milligrams)	Dry Weight After Three Months (in milligrams)	Change (before minus after)
Sample plastic—1	134	122	12
Sample plastic—2	143	117	26
Sample plastic—3	129	119	10
Sample plastic—4	137	125	12
Sample plastic—5	147	136	11
Total Weight: Plastic	690	619	71*
Sample plastic—1	339	223	116
Sample plastic—2	347	221	126
Sample plastic—3	376	211	165
Sample plastic—4)	388	233	155
Sample plastic—5	355	249	106
Total Weight: Paper	1,805	1,137	668*

*In these cells, Connor found he could check his arithmetic. The rows yield the same answer as the columns.

The mean or the average

Connor's totals showed him the change in all five samples, but the problem of variation still bothered him. All his samples started out weighing something different and ended up different, too. Could he find one number that would get rid of individual variation and show him a general trend?

His solution was to calculate the **mean** (also called the **average**). *The mean is one number that varies from all the other numbers by the least amount. It is calculated by adding all the observations together and dividing by the total number of observations.* Connor had five cases of each kind of sample, so he totaled his weights and divided by five in each instance.

For example, the mean weight of plastic bags before burial equals

$$\frac{134 + 143 + 129 + 137 + 147}{5} = 138 \text{ milligrams}$$

Connor made a new table:

Material	Mean Dry Weight Before Burying (in milligrams)	Mean Dry Weight After Three Months (in milligrams)	Mean Change (before minus after)
Plastic	138	124	14
Paper	361	227	134

Rounding and significant figures

Notice that Connor used only whole numbers in his table, no decimals, even though not all his means turned out to be whole numbers. He rounded his means to the nearest whole number, using

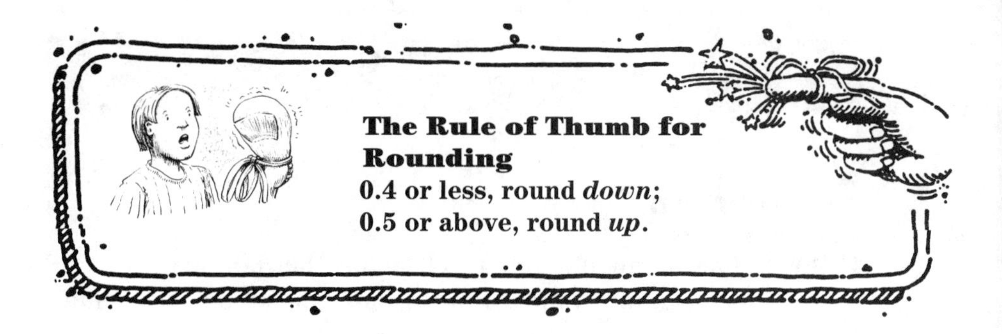

The Rule of Thumb for Rounding
0.4 or less, round *down*;
0.5 or above, round *up*.

For example, the mean weight of his paper samples after three months was

$$\frac{223 + 221 + 211 + 233 + 249}{5} = \frac{1137}{5} = 227.4 \text{ rounded down to } 227$$

and

the mean weight of his plastic samples after three months was

$$\frac{122 + 117 + 119 + 125 + 136}{5} = \frac{619}{5} = 123.8 \text{ rounded up to } 124$$

Why didn't Connor use the decimals? They sound more exact, don't they? That's precisely why Connor didn't use them. He knew that his calculations could not be more precise than his measurements. He had weighed to the nearest milligram. His average could not be any more exact than milligrams.

Scientists call this rule **significant figures**. Your calculations cannot have more significant figures than your original measurements.

Some additional examples:

Original measurements	Mean: WRONG number of significant figures	Mean: CORRECT number of significant figures
47.3, 51.5, and 63 cm	53.93 or 54 cm	53.9 cm
1975, 1983, and 1999 gm	1985.6, 1985.67, or 1985.667 gm	1986 gm or 1.986 kg
98.6°C, 99.5°C, and 101.7°C	99, 99.0, 99.93, or 99.933°C	99.9°C

The median

Connor's measurements were all close together, so the mean served him well. Sometimes, however, a mean can be misleading, especially when one or a few measurements vary greatly from all the others. In that case, it may be better to use the **median**.

The *median is the middle number when all numbers are arranged in order* from largest to smallest or from smallest to largest. For example, the mean of 22, 26, 31, 27, and 125 is 46. That number doesn't represent any of the findings very well. Assuming that the large value really is correct (it looks suspicious, doesn't it?) the median might better summarize the data:

125
31
27 = median
26
22

The table includes all the numbers in the data set, repeated as many times as they actually occur. For an even set of numbers, average the middle two values to find the median:

125
31
31
27
$(27 + 25) \div 2 = 26$ The median is 26.
25
22
22
21

The mode

Sometimes neither the mean nor the median best describes all the data. In that case, the **mode**—the most frequently occurring number—may be the best choice. For example, in the following set, 6 occurs so many times it deserves attention.

19, 18, 16, 12, 10, 9, 8, 8, 6, 6, 6, 6, 6, 3, 1

Mean = 9
Median = 8
Mode = 6

Percent change

By using the mean, Connor smoothed out the variations of his individual measurements. Making comparisons, however, was still a problem. An average change of 14 milligrams compared with 134 milligrams looked important, but *how* important?

Connor realized that he needed to compare his means against a common standard. He wondered, "What if the bags had all weighed the same in the beginning?" Connor asked his math teacher for advice and learned that calculating a percent is an easy way to compare numbers. *Percent compares all numbers on the basis of 100 units.*

Percent answers the question:
Out of every 100 parts, how many are ________?

Percent change gave Connor a way to compare the before and after weights based on a common unit of 100. In his case, the unit was milligrams. Percent change, therefore, gave a summary of how much change occurred in every 100 milligrams of original weight.

Connor calculated percent change as the difference between the before and after weights (before minus after) and divided by the original weight.

Thus, for his plastic bags, on the average, the percent change was

$$14 \div 138 = 10\%$$

The percent change of his paper bags was

$$134 \div 361 = 37\%$$

NUMBER CRUNCHING II: ADVANCED TECHNIQUES

Attention Middle School Students
You may not need the techniques summarized in this section right now. Maybe you'll use them later—in high school. For now, you may want to skip to "Graphs" beginning on page 169. Your teacher will tell you how much number crunching you need to do.

Percent change gave Connor something he could work with. His paper bags had lost almost four times more weight than his plastic bags. That looked like a major and important difference to Connor. Would a scientist agree?

Maybe. Maybe not.

Scientists know that things can vary, or change, just by chance. Although the difference Connor found looks too big to be an accident, scientists don't rely simply on hunches to decide which numbers represent real differences and which can pop up randomly, by chance.

Significance and probability

In many cases, chance can be calculated as a probability. Probability calculations answer the question "How likely is (some event or occurrence)?"

For example, think about the probabilities you hear on the weather report:

- "The chance of rain today is 10 percent." You can risk leaving your umbrella at home and run only 10 chances in 100 (or 1 chance in 10) of getting wet.

- "The chance of rain today is 90 percent." If you forget your umbrella, you're pretty sure to get drenched.
- "The chance of rain today is 50 percent." Should you take your umbrella? You might as well toss a coin to decide. Your decision may be based more on your personality than on the weather prediction.

In popular newspapers and magazines, you'll often run into probabilities as measures of risk to your life and health.

For example, which is the more likely cause of accidental death in the United States—fire, drowning, a fall, gunshot, or poisoning? If you guessed a fall, you know your probabilities. On the average, the chances are*

Fall	1 in 20,000
Poison	1 in 33,000
Drowning	1 in 50,000
Fire	1 in 50,000
Accidental gunshot	1 in 167,000

These probabilities are misleading, however, because they assume that everyone has an equal chance of falling or getting shot. Not so. Hunters are at a greater risk of accidental gunshot wounds, and mountain climbers increase their risk of falling considerably. Some increased risks are beyond our control. The elderly are more likely to die from a fall than the young.

Still, as unlikely as death by accidental gunshot may seem, it's still far more probable than being struck by lightning. Overall, in the United States, an individual's chance of being hit by a bolt from the blue is about 1 in 600,000. However, where you live makes a difference. Injury from lightning strikes occurs nearly twice as often in Florida as in Michigan; and a Florida resident is 16 times more likely to die from a lightning strike than is a Californian resident.

*Calculations are based on total population and death rates, last census year 1990.

Scientists use estimates of probability to draw conclusions about the effect of any change they make (or nature makes) in the independent variable. Tests of significance answer the question, "Did the change in the dependent variable result from change in the independent variable or occur just by chance?"

Some statistics that help answer this question include *standard deviation of the sample*, the *t-test*, and the *correlation coefficient.* Many others are available and may be appropriate, depending on the nature of your project.

The formulas for these statistical tests can be quite complex. Fortunately, calculators and computers will do the arithmetic for you. All you need to understand is what the numbers mean.

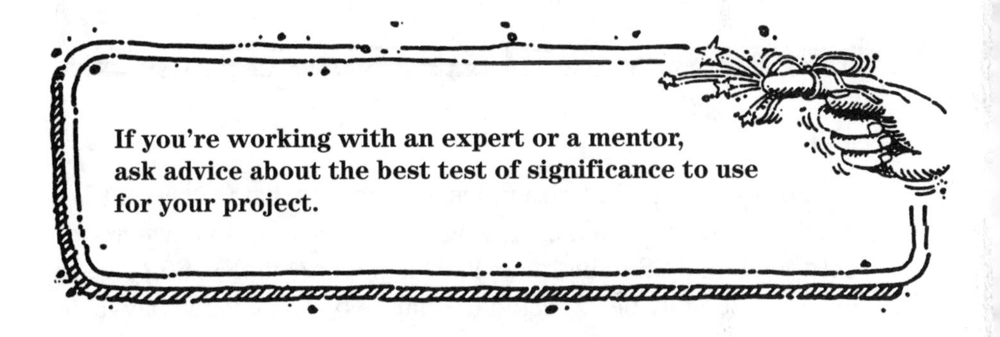
If you're working with an expert or a mentor, ask advice about the best test of significance to use for your project.

Standard deviation

For most sets of measurements, you can appropriately calculate the *standard deviation (S.D.)* of the sample.* Many handheld calculators and computers will figure S.D. for you. What does the S.D. mean once you have it?

Standard deviation is an expression of the variation in a sample. If every member of a sample were exactly the same, the S.D. would be zero (0). However, nature seldom works that way. Some individual measurements vary greatly from the mean; others vary only a little. S.D. answers the question, "What is the average amount of variation from the mean?" A big number

*Calculation of the S.D. of the population is not possible, because only a sample from the population was studied.

signals big variations. A small number says most of the individual measurements fall close to the mean.

Connor calculated standard deviations for the changes in weight of plastic and paper bags. He found that the S.D. of his plastic samples equaled 7. The S.D. of his paper samples equaled 25.

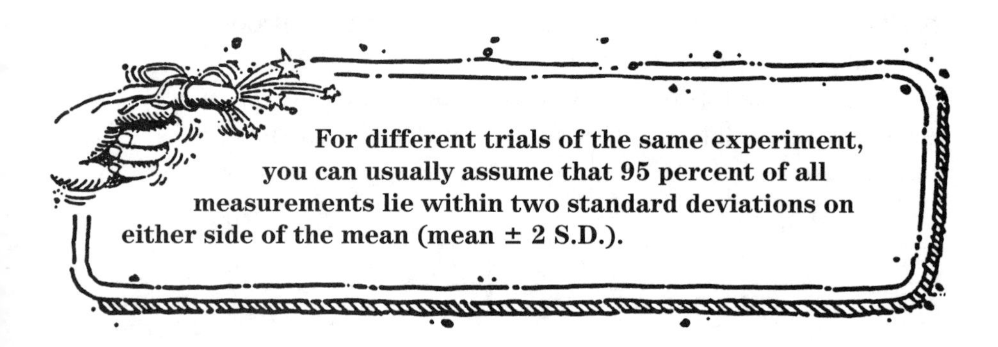

For Connor's plastic samples, that meant that 95 percent of all the bags he could ever weigh would change somewhere between 0 and 28 milligrams. For his paper samples, 95 percent could reasonably be expected to change by some value between 84 and 184 milligrams.

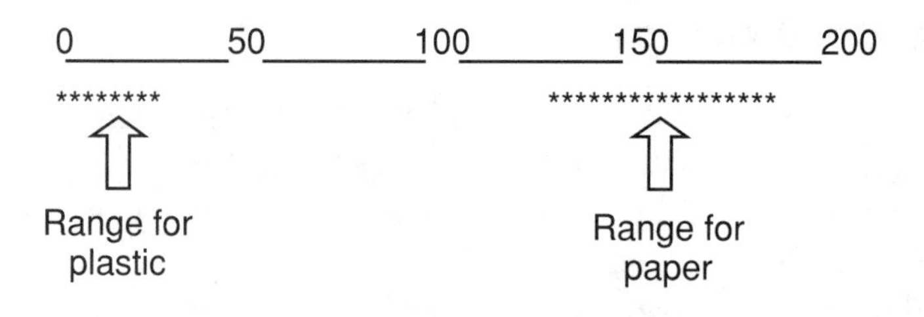

These two ranges come nowhere near overlapping; therefore, Connor felt confident that the changes truly depended on the material the bags were made from and did not result from some chance variation.

The *t*-test

Connor's math teacher helped him take his analysis a step farther.

"A *t*-test," Mrs. Porter said, "will give you a probability of whether these differences are real or just chance."

Mrs. Porter logged on to the Internet and went straight to the statistics Web site maintained by the University of California, Los Angeles.* "There are many other places where we could do the same calculations," she said, "but I find this one quick and easy to use."

Mrs. Porter showed Connor how to enter his numbers. After a few minutes, the UCLA calculator gave Connor this information:

> *t*-statistic: 10.182466
> degrees of freedom: 4.547083
> *p*-value: 0.000266

"What do all those numbers mean?" Connor asked.

"The *t*-statistic and the degrees of freedom mean a lot to statisticians," Mrs. Porter replied, "but for you the *p*-value is the heart of the matter."

Connor studied the number. "I could round it to three ten-thousandths [0.0003]," he said.

His teacher beamed. "You got it!" she exclaimed. "That's exactly what the *p*-value means. *P* stands for probability. The number is three ten-thousandths [0.0003]. Chances are 3 in 10,000 that your changes in weight happened just by chance."

"That's not very likely, is it?" Connor asked, grinning.

"It's very unlikely indeed," she answered. "In fact, scientists are usually satisfied with far smaller differences. They report their probabilities as (she wrote on the chalkboard):

- $p < 0.05$ (less than 5 chances in 100);
- $p < 0.01$ (less than 1 chance in a 100); or
- $p < 0.001$ (less than 1 chance in 1,000).

"Your p value is less than 1 in 1,000, so you can report your results as highly significant," Mrs. Porter concluded. Connor nodded. Statistics, he thought, really aren't that hard after all.

**http://www.stat.ucla.edu/calculators*

Correlation

Vanessa's project differed from Connor's. She wanted to see if study time affected scores on a simple math quiz. Her school's human subjects committee approved her research, and she recruited 12 ninth-grade students at random to take part. She divided her subjects into groups of three and gave them study times ranging from 5 to 20 minutes. Then each student took the quiz. Vanessa completed her data table with percent correct scores for each student.

Student	Study Time (in minutes)	Percent Correct on Quiz
A	5	64
B	5	92
C	5	61
D	10	75
E	10	76
F	10	81
G	15	85
H	15	67
I	15	91
J	20	83
K	20	95
L	20	98

When she examined her data, Vanessa thought she saw a pattern, but she couldn't be sure. Some inconsistencies bothered her. Student B did almost as well as students who studied four times longer, and H did little better than two students who studied for only five minutes. She wondered if her data revealed anything important.

Connor told Vanessa about Mrs. Porter and her statistics Web site, so Vanessa dropped in for a visit.

"A correlation coefficient will help you," Mrs. Porter said as she logged on to the Internet, just as she had for Connor.

"A cory-what?" asked Vanessa.

Mrs. Porter smiled. "A correlation coefficient tells you if two sets of numbers are related to each other in any consistent way.

This number (often represented by the symbol r) varies between +1 (plus 1) and –1 (minus 1). A zero means no correlation. The numbers don't vary together in any consistent way. R–values above 0 and approaching +1 signal a *positive correlation*. That means the two sets of numbers rise or fall together. As one gets bigger or smaller, so does the other. A number below zero approaching –1 means a *negative correlation*. As one number gets bigger, the other gets smaller. The bigger the r–value—either positive or negative—the stronger the relationship."

Mrs. Porter showed Vanessa how to enter her data table into the computer. Soon, a value of r = .59 appeared.

"That means a positive relationship," Vanessa said. "Studying more really does improve scores, despite the differences of a few students."

"That true," Mrs. Porter said, "and my guess is that if you test a larger number of students, you'll get an even stronger correlation. Increasing sample size reduces variation, you know."

Vanessa nodded, distracted. She was studying the graph just coming out of the printer.

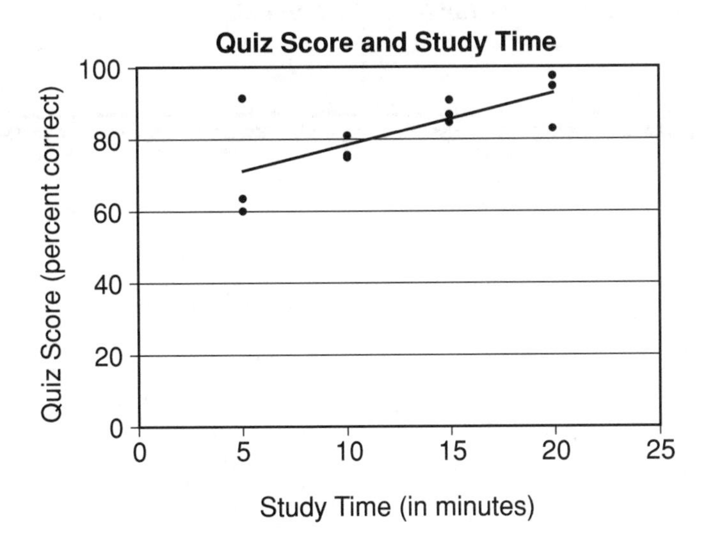

"The points are the students' study times and scores," Mrs. Porter said. "The line is called a regression line. It shows the general trend in your data."

"The line slants upward," Vanessa said.

"That upward slant confirms the positive correlation," Mrs. Porter explained. "As one number gets bigger, so does the other, so the line slants upward. If the line slanted down, you'd have a negative correlation with your dependent variable decreasing as your independent increased. Get it?"

"I think so," Vanessa said, taping the graph into her journal and making a note to learn more about statistics.

Do you—like Vanessa—want to learn more about statistics? First, make friends with your math teacher. Second, visit the library to borrow copies of *The Cartoon Guide to Statistics* by Larry Gonick and Woolcott Smith (NY: HarperCollins, 1993) and *Statistics Without Tears: A Primer for Non-Mathematicians* by Derek Rowntree (NY: Prentice Hall, 1981).

GRAPHS

As Connor and Vanessa learned, tables, calculations, and statistics can help you make sense of your data and reveal what all those numbers mean. An even more powerful tool for getting meaning from data is the **graph**. *Graphs draw pictures of data. They make comparisons visual, immediate, and concrete*. Getting meaning from tables can be hard work; however, a quick glance at a graph is all that's needed for a trend or pattern to leap off the page at you.

Getting meaning from graphs is quick and easy for you and for others who will see your work, as long as you use the right kinds of graphs and construct them correctly.

Kinds of graphs

Many different kinds of graphs are commonly used to show relationships between and among sets of numbers. The three most often used for science projects are

- bar graphs, for example:

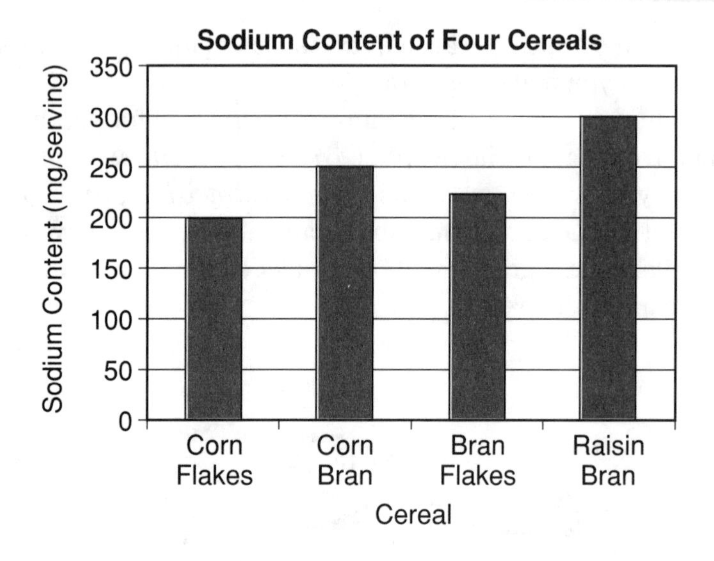

- line graphs, for example:

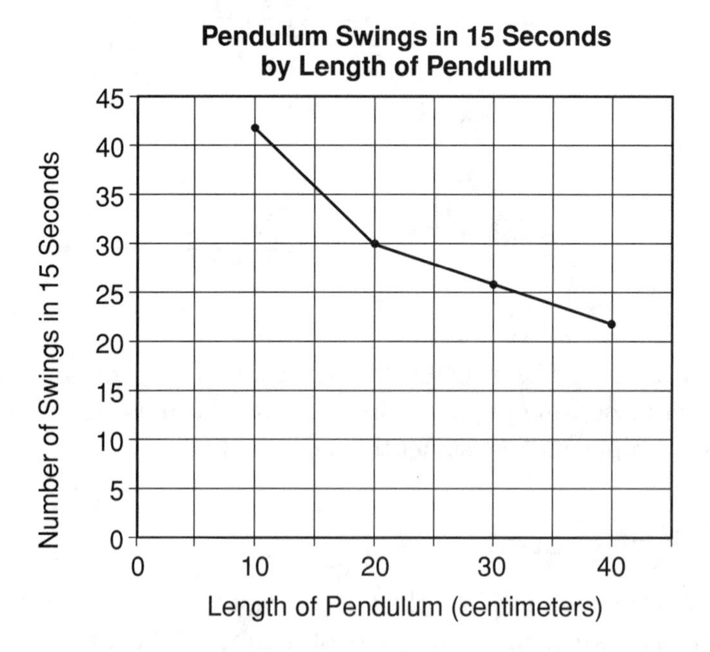

- pie or circle graphs, for example:

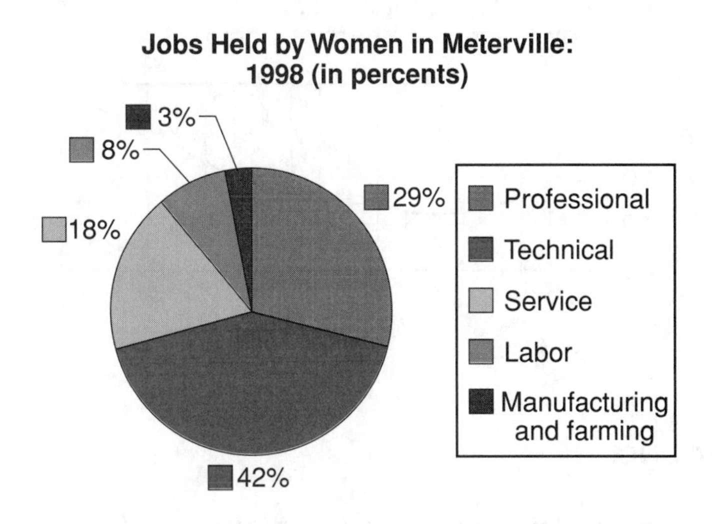

How do you know which kind of graph to use? The answer depends on the nature of your data.

- On a line or bar graph, the *dependent* variable (outcome) always goes on the *vertical* (y) axis.
- The *independent* variable (that changes naturally or the experimenter changes deliberately) always goes on the *horizontal* (x) axis.

RIGHT:

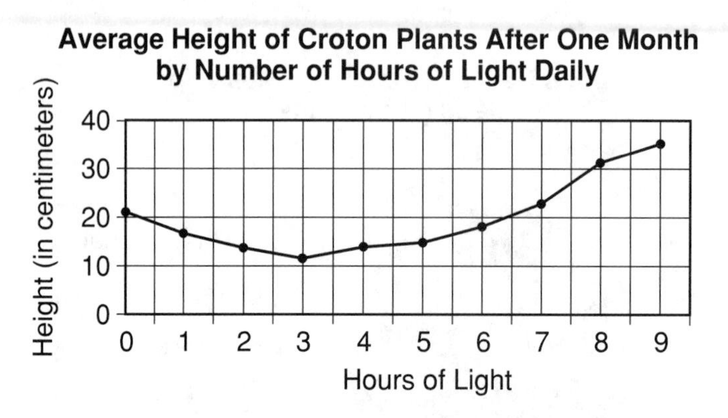

WRONG:

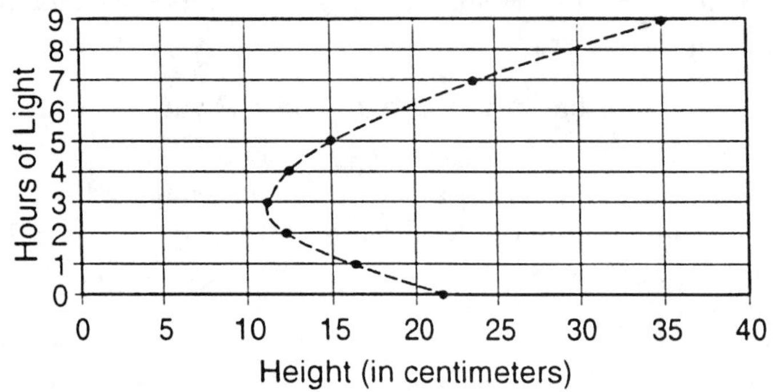

Your choice of a bar graph or a line graph depends on whether the independent variable is *continuous* or *discontinuous*.

Continuous variables

Continuous variables change in a fixed, unchangeable way. They have a natural order. Examples include time, weight, height, distance, speed, and so on. Continuous variables

- are always numbers;
- occur in a fixed order (that is, their sequence cannot be changed);
- begin with 0 and may extend as far as desired into either positive (or sometimes negative) numbers; and
- are best suited to line graphs.

Discontinuous variables

Discontinuous variables have no natural or numerical order. You may arrange them in any order you wish. Discontinuous variables fall into categories such as boys and girls, positions on a football team, flavors of milkshake, and votes in an election.

Discontinuous variables

- are either numbers or words (but usually words);
- are discrete and separate (that is, one can be left out without affecting the others);
- can be arranged in any order;
- make good bar graphs, but should **never** be used in line graphs; and
- make good pie graphs **if and only if** the categories represent parts of a whole (and total 100 percent of something).

Some variables may be either continuous or discontinuous, depending on how you handle them. Numbers in natural numerical order are continuous; however, if you group them in ranges or categories, they become discontinuous. For example, the ages of respondents in a survey are a continuous variable if all ages are represented and *not* categorized. Age becomes a discontinuous variable if you lump the respondents into groups, for example: 5 years or younger; 6 to 10; and 11 to 20.

The distinction between continuous and discontinuous variables tells you what kind of graph to make. In general,

- *Discontinuous independent variables make bar graphs.*
- *Continuous independent variables make line graphs.*

For example, a single serving of one popular brand of corn flakes with one-half cup skim milk provides:

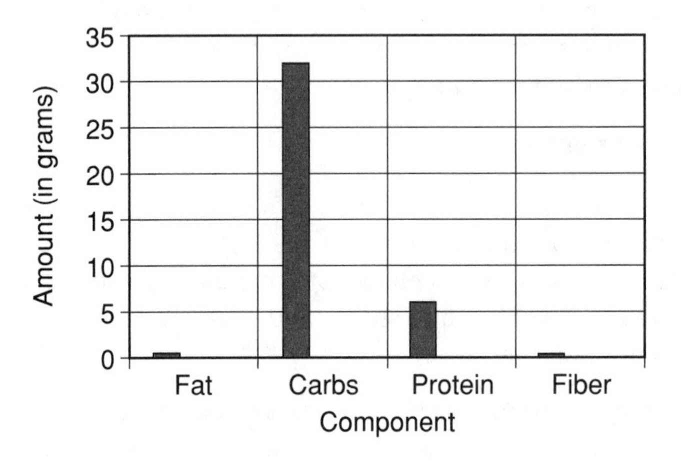

Study the example above. The dependent variable is the amount in grams. It is a continuous variable. It is plotted on the y-axis. It has a natural order (increasing from zero) that cannot change.

The independent variable on the x-axis is a discontinuous variable. The experimenter selected these categories of nutrients and might have chosen others. These categories have no natural order. For ease of reading and explanation, you might choose to arrange such variables in ascending order (from least to most) or descending order (most to least). For example, the same information as above might be shown as

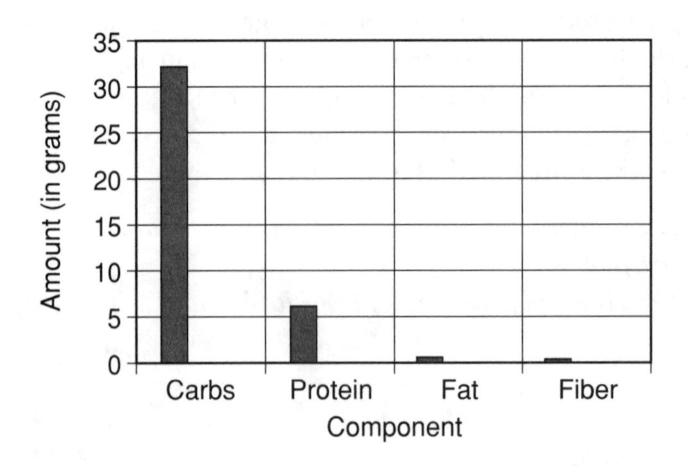

or

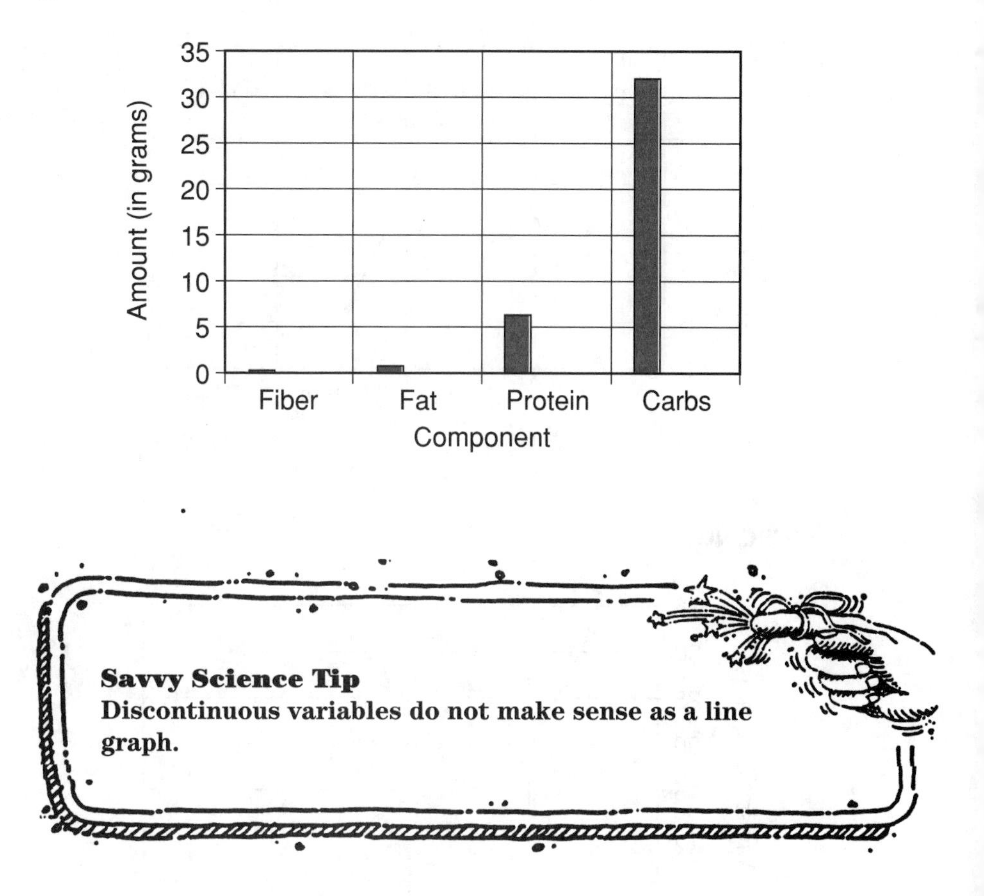

Savvy Science Tip
Discontinuous variables do not make sense as a line graph.

Notice that the cereal data do *not* make sense as a line graph. A line shows a trend in a continuous dependent variable. A line is deceiving when its shape can be changed simply by changing the order of categories of independent variable. For example,

WRONG:

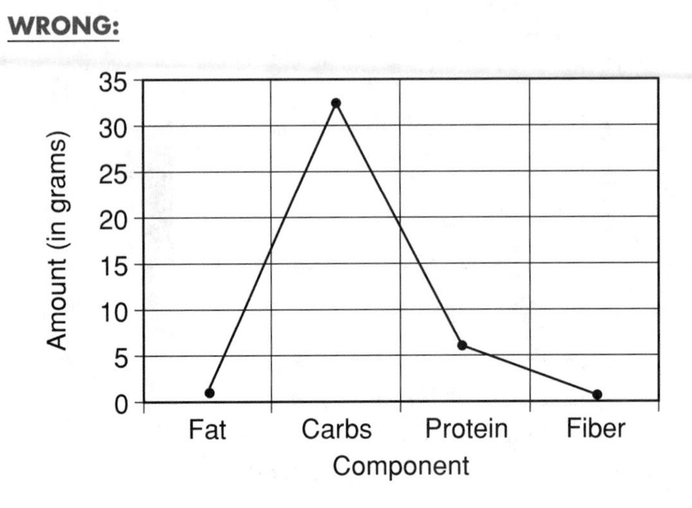
Amount (in grams)
35
30
25
20
15
10
5
0
Fat
Carbs
Protein
Fiber
Component

WRONG:

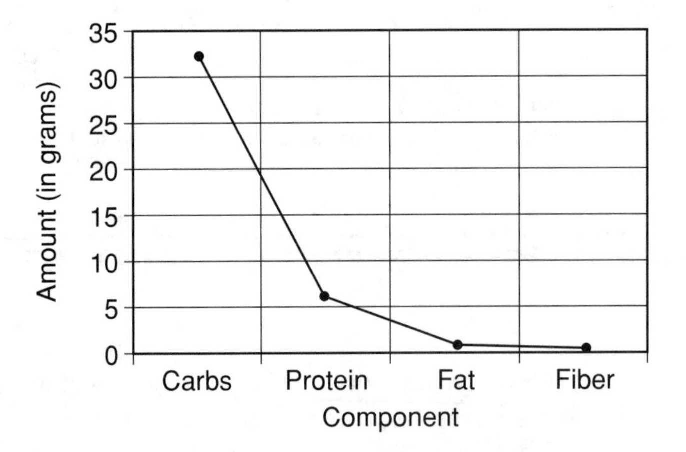
Amount (in grams)
35
30
25
20
15
10
5
0
Carbs
Protein
Fat
Fiber
Component

WRONG:

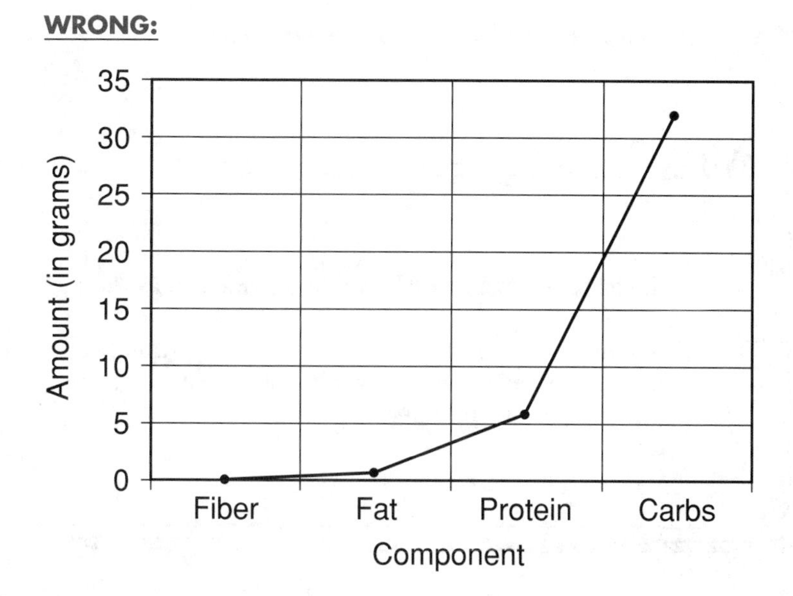

Neither would these data make sense as a pie or circle graph, since the experimenter is not showing the parts of a whole. (A serving of cereal contains more than these four parts.)

WRONG:

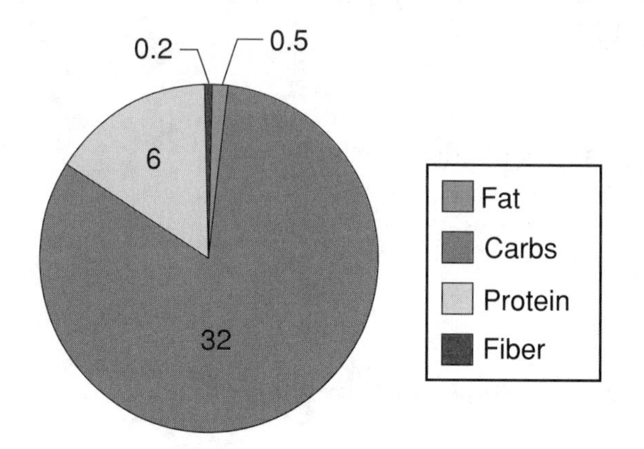

How to make a line or bar graph

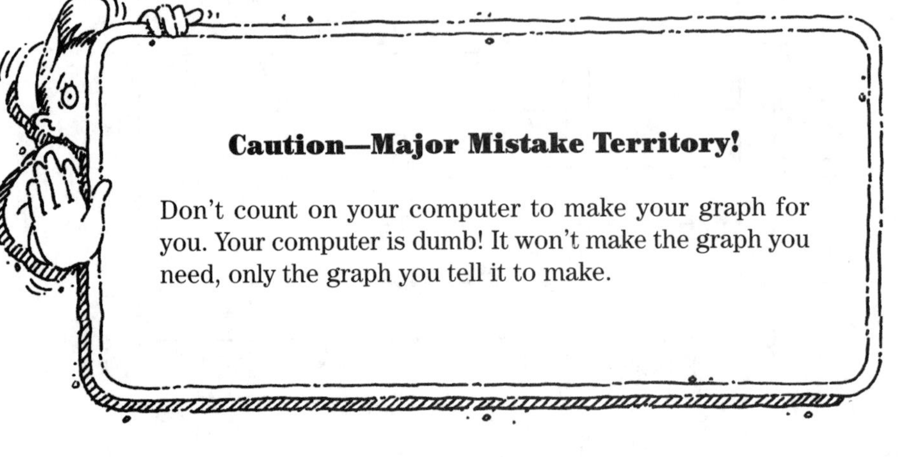

Steps:

1. Recall which is your dependent variable and which is your independent variable. On graph paper, draw a line to represent the vertical axis (y) along the left side. Draw a line for the horizontal axis (x) from left to right along the bottom of the page.

y axis = dependent variable = ⇧

x axis = independent variable =

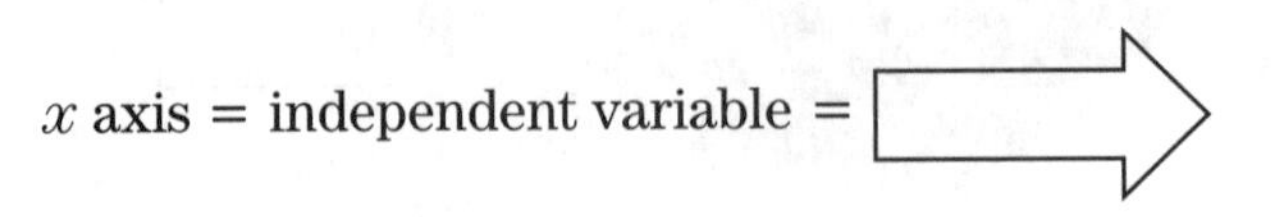

2. For continuous variables (such as time, weight, and distance), mark off the x-axis as a number line that begins with 0 and goes far enough to include your largest number. Use whatever equal divisions are convenient and fit on the paper.

   For example, for data values 3, 5, 6, 7, 9, 14, 22, 24, a number line from 0 to 30 divided into increments of five should do fine:

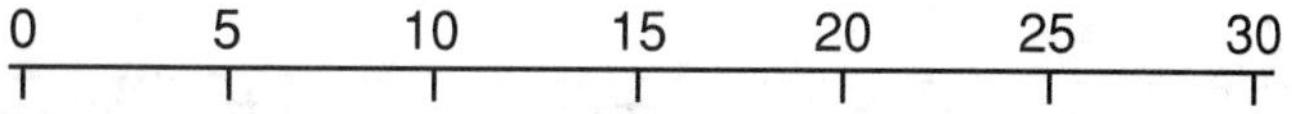

```
0       5       10       15       20       25       30
|-------|-------|--------|--------|--------|--------|
```

   The points fit along the line, like this:

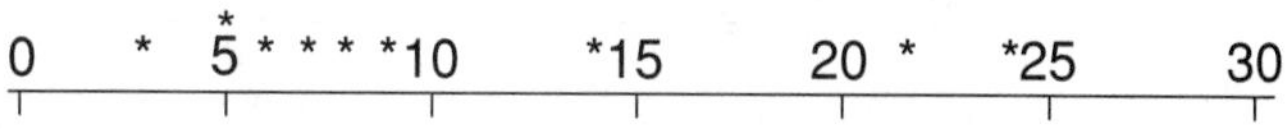

```
            *
0     *     5 * * * *10       *15       20  *    *25       30
|-----------|-----------|---------|---------|---------|---------|
```

   For bigger numbers, make small spaces represent larger units.

   For example, for data values 175, 169, 277, 386, 344, a number line from 0 to 400 divided into increments of 50 works well:

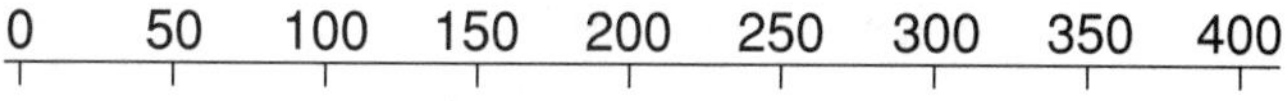

```
0     50    100    150    200    250    300    350    400
|------|------|------|------|------|------|------|------|
```

   The data points can be placed:

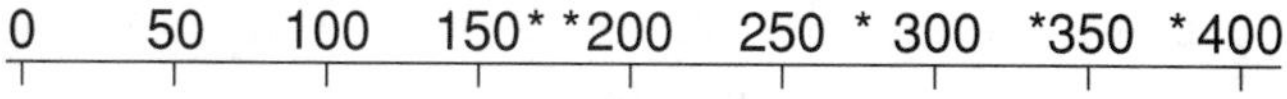

```
0     50    100    150* *200    250 * 300   *350  *400
|------|------|------|------|------|------|------|------|
```

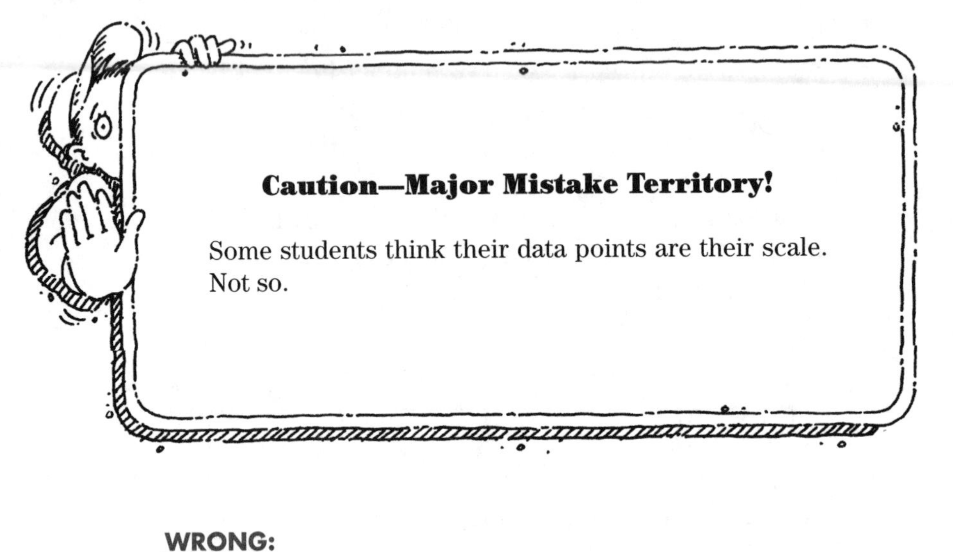

Caution—Major Mistake Territory!

Some students think their data points are their scale. Not so.

WRONG:

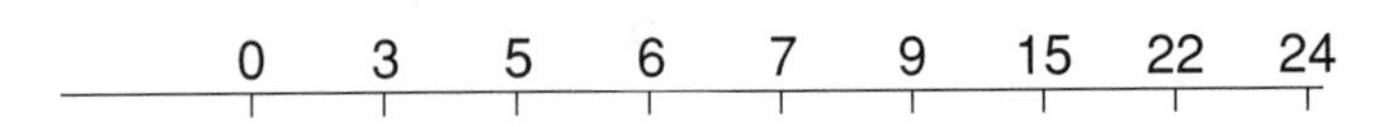

Also, some forget that equal spaces must be used to represent equal numbers.

WRONG:

0 5 10 15 20 25

Some forget to plan ahead:

WRONG:

0 5 10 15 20 25 30 35 40 50 100

Don't number the spaces. Number the lines. The space between the lines represents your unit. For example, the spaces between the lines in the following graph represent ten units.

RIGHT:

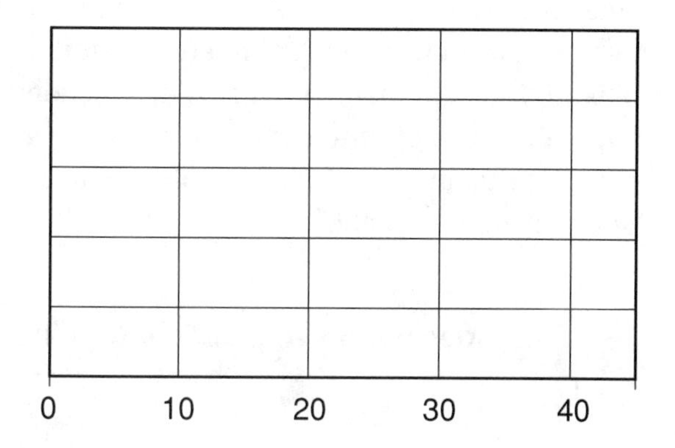

WRONG:

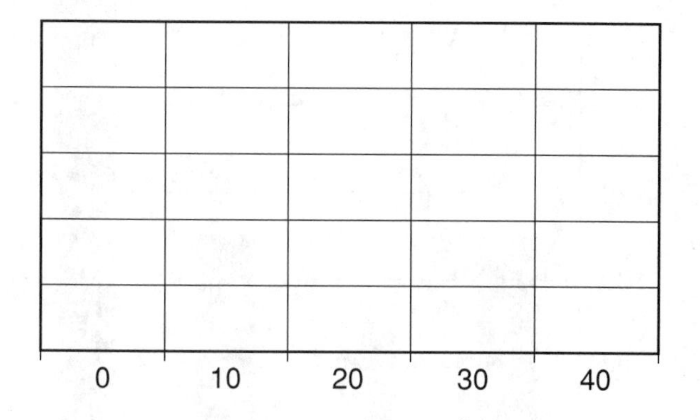

Notice that with the lines numbered, you can easily find where a number like 16 or 23 or 38 goes. With the spaces numbered, it's impossible to tell where to put these or any other numbers.

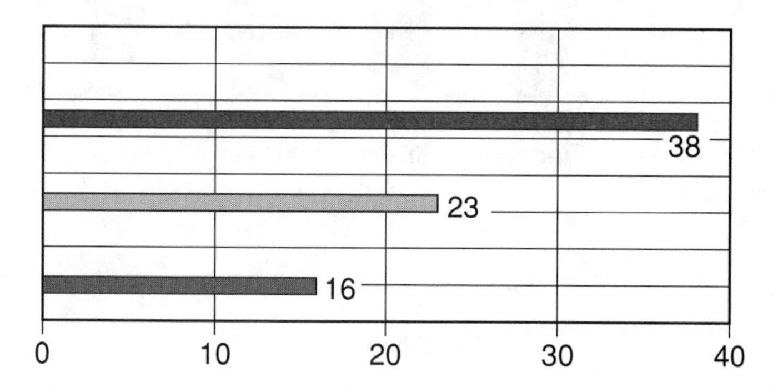

Do the scales on the x-axis and y-axis have to be the same? No. Make them to accommodate whatever you have measured as your dependent and independent variables. If your independent variable is discontinuous, simply measure off equal spaces for each category. For example, this independent variable has four categories of the experimenter's choice:

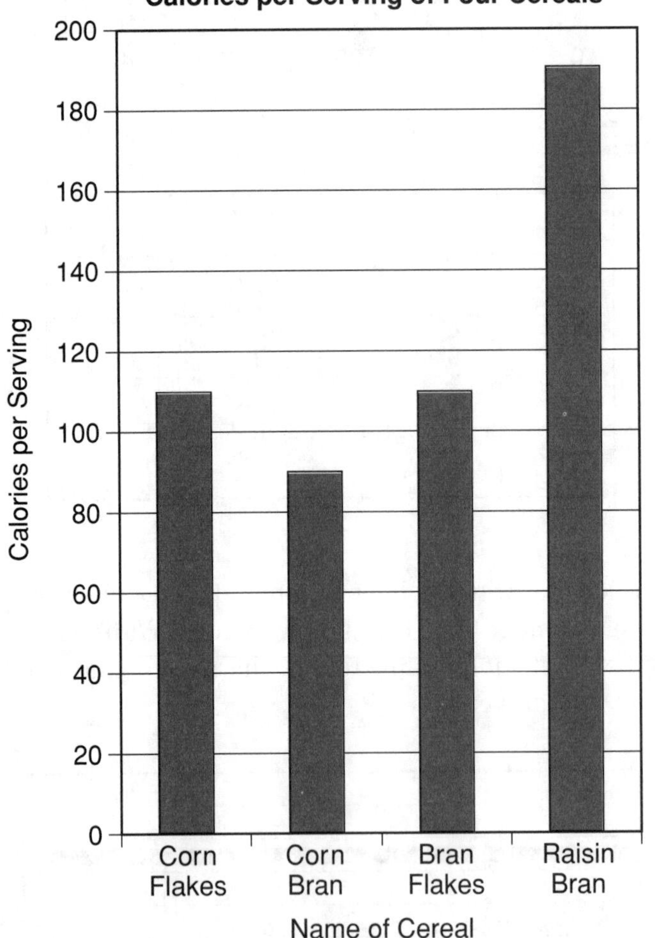

Always start your number lines at zero. Why? Because any other scale misrepresents the relationship between numbers. Scales without a zero point can make small differences look very large, which can mislead both the experimenter and people who see or read about the project.

For example, for data values 315, 346, 355, 375, the difference between these numbers looks small if you use a scale of 1 to 400. That represents their true relationship.

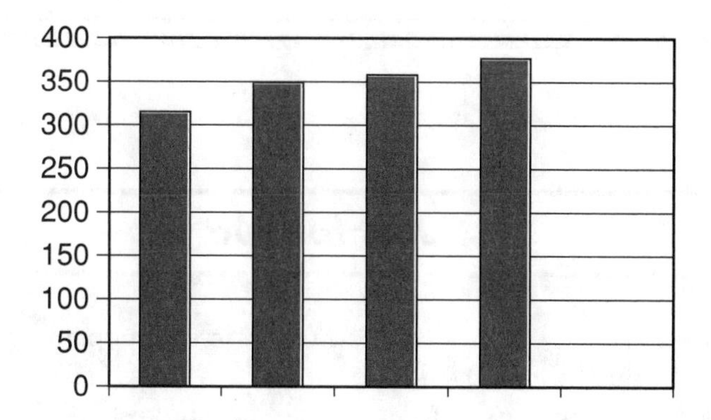

A scale that runs from 280 to 380, though, falsely exaggerates the differences. Notice how the last bar in the next figure looks more than double the size of the first. Such scales misrepresent the relationships between numbers and mislead the graph reader.

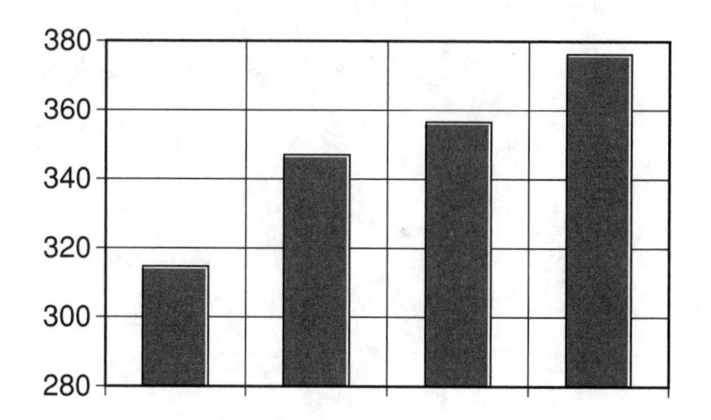

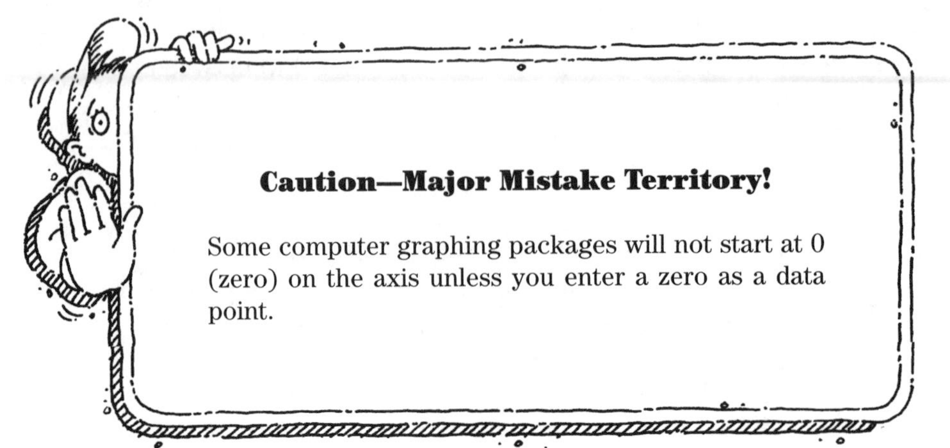

Caution—Major Mistake Territory!

Some computer graphing packages will not start at 0 (zero) on the axis unless you enter a zero as a data point.

Just for Fun

Look at the way the business section of the newspaper sometimes graphs stock prices. By leaving zero off the scale, these graphs make small changes in trading prices look large. That's fine for investors for whom a fraction of a cent can represent a fortune. However, it does not represent the true change in the stock's value.

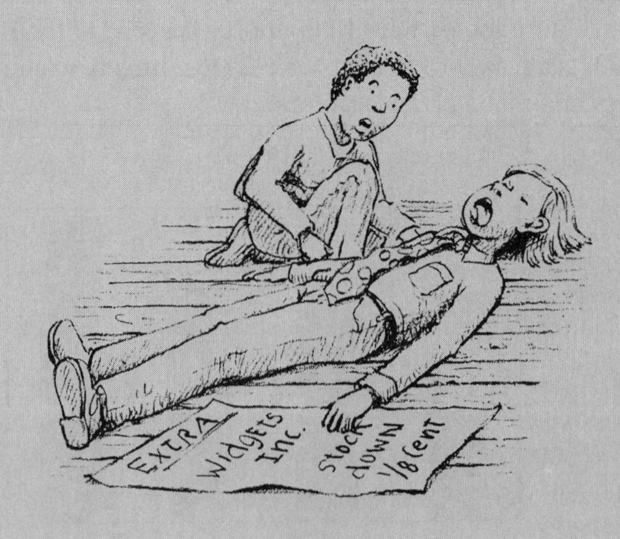

3. Decide on groupings for independent variables if you need to. If your independent variable is discontinuous (such as kinds of cereal or makes of cars), then your data are grouped already. Some continuous data fall into natural groupings, too.

For example, Lisa collected the measurements of hat sizes taken for all the graduating seniors at Spellman High School when they ordered their caps and gowns. Pages and pages of records gave her the sizes for all 198 seniors. She realized that using all 198 points on her graph was unnecessary. Instead, she grouped her data according to the number of students ordering each hat size. Her data table looked like this:

Hat Size	Number of Seniors
$6\frac{1}{2}$	3
$6\frac{5}{8}$	12
$6\frac{3}{4}$	23
$6\frac{7}{8}$	45
7	52
$7\frac{1}{8}$	41
$7\frac{1}{4}$	14
$7\frac{3}{8}$	6
$7\frac{1}{2}$	2
$7\frac{5}{8}$	0
Total	198

Lisa realized that she could leave her data the way they were or group them even further. She knew she could easily use the size ranges provided by the cap and gown rental company: extra small ($6\frac{1}{2}$–$6\frac{5}{8}$); small ($6\frac{3}{4}$–$6\frac{7}{8}$); medium (7–$7\frac{1}{8}$); large ($7\frac{1}{4}$–$7\frac{3}{8}$); and extra large ($7\frac{1}{2}$–$7\frac{5}{8}$).

Hat sizes are already divided into even categories (by $\frac{1}{8}$) and grouped by twos (into extra small, small, and so on). You cannot group some data so easily.

For some series, you will need to decide what the logical groupings are. You have a lot of freedom in doing that, *as long as the groups encompass an equal range*. For example, Lisa wanted to see if foot length correlates with head size. She measured the feet of all 198 Spellman High School seniors and got lengths ranging from 21 centimeters to 37 centimeters Most students measured somewhere between 25 centimeters and 29 centimers.

Which of the following sets should Lisa use to categorize length on her graph? Why?

Set A	Set B	Set C
21–24	21–23	21–24
25–28	24–27	25–29
29–32	28–29	30–33
33–36	30–36	34–36

Set A is the best choice. To see why, look at the lengths of number lines for these sets.

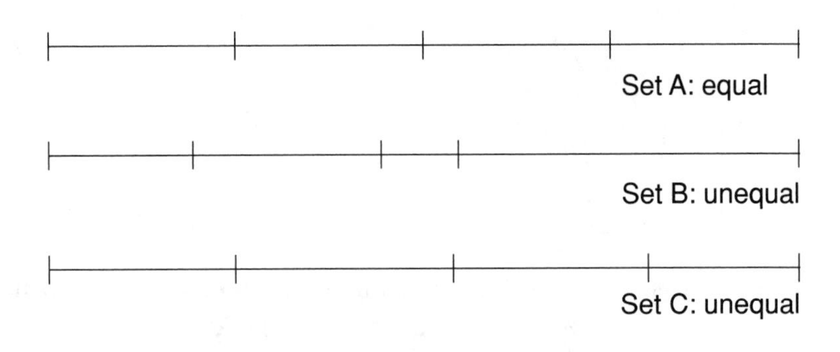

In set A, the groups all cover the same range (4 centimeters), and the smallest and largest numbers fit. Group sizes in sets B and C are unequal. Set C artificially groups the most frequent measurements together. Such purposeful engineering of data can lead to erroneous conclusions.

It is often better to use more groups rather than fewer. Lisa's data will show greater detail if she breaks the groups in set A down to 21–22, 23–24, 25–26, and so on.

4. The next step is to *plot points*. That means finding the point on the graph where the x-value and the y-value meet from their respective number lines.*

   Look back at Lisa's original table of hat sizes. Three students ordered a size $6\frac{1}{2}$. To plot that point, Lisa makes a dot where the hat size $6\frac{1}{2}$ meets the number of students, three. Then she moves on to the next pair and makes another dot. After she makes all her dots, her graph looks like this:

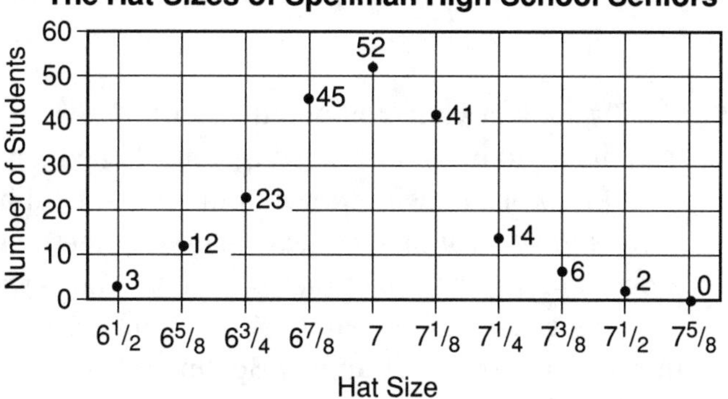

5. Once you plot all points, connect them with a line for continuous variables or make them into bars for discontinuous variables. For example:

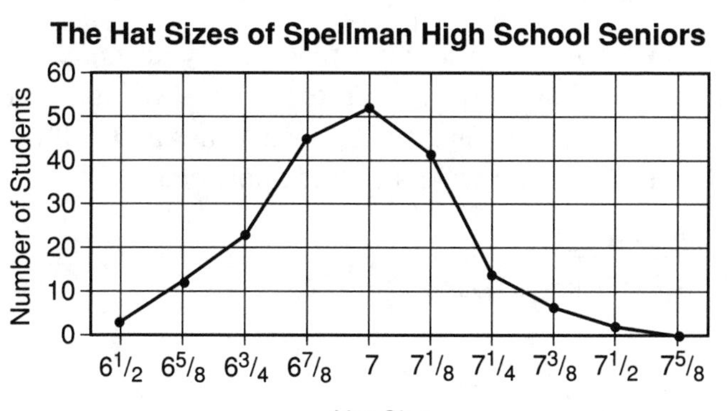

*Alas, this is one of those rare cases when it's OK to leave zero off the x-axis. Since no hat size smaller than 6 exists, it acts as the zero point.

or

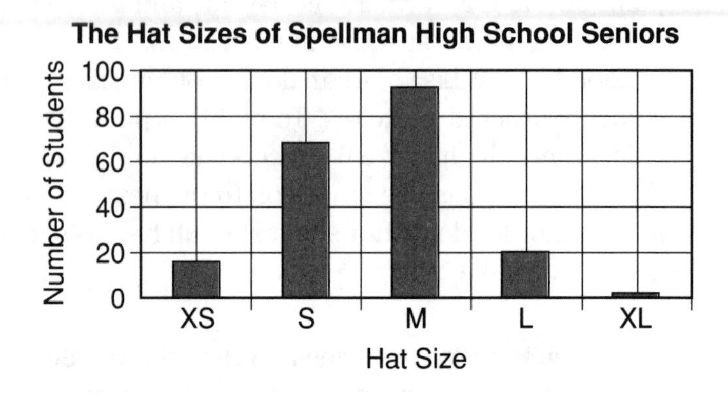

6. Finally, look at your graph and *read its meaning*. Your graph should make interpreting your data much easier than just working with tables of numbers. Graphs tell stories. As one thing changes, so does something else in a predictable—or sometimes, unpredictable—way.

   Lisa's graphs immediately show which hat sizes are more or less frequent among Spellman High School seniors. Can she generalize to all seniors in the country? Perhaps, since she has no reason to believe that Spellman seniors have significantly bigger or smaller heads than seniors at any other school.

The key to graphing: Relationships

Remember that your graph is a picture of your numbers. That picture must show how one number relates to another. For example, if one number is three times another and half of still another, those relationships should show clearly.

FOR EXAMPLE:

12 is 3 times 4 and $\frac{1}{2}$ of 24.

Notice how the size of the bars reveals that relationship in this graph:

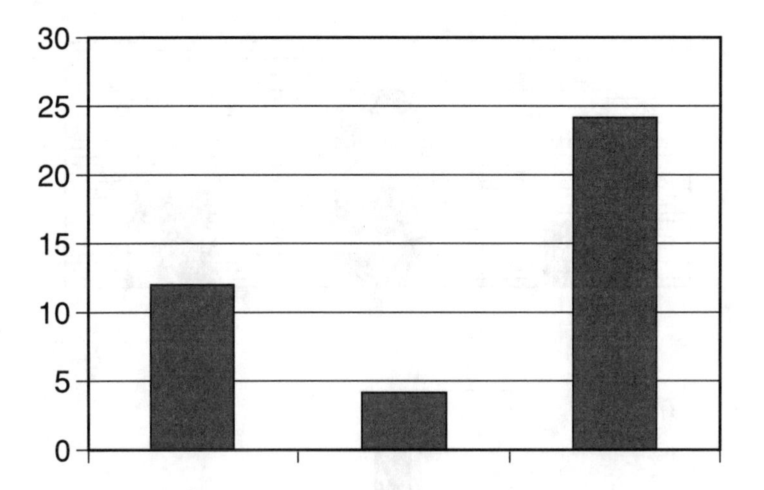

What is the relationship among these numbers? (Use a ruler if you need to.)

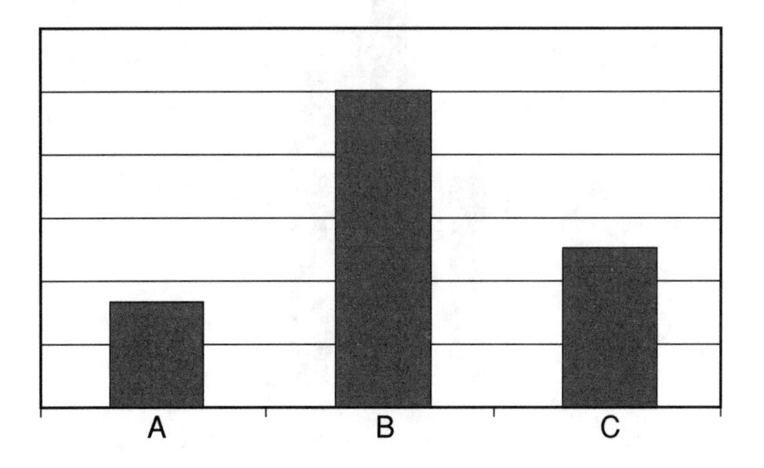

All the following statements are equally correct:

- B is twice C and 3 times A,
- C is $1\frac{1}{2}$ times A and $\frac{1}{2}$ of B, and
- A is $\frac{1}{3}$ of B and $\frac{2}{3}$ of C.

Notice that the relationship does not change no matter how the size or scale of the graph changes.

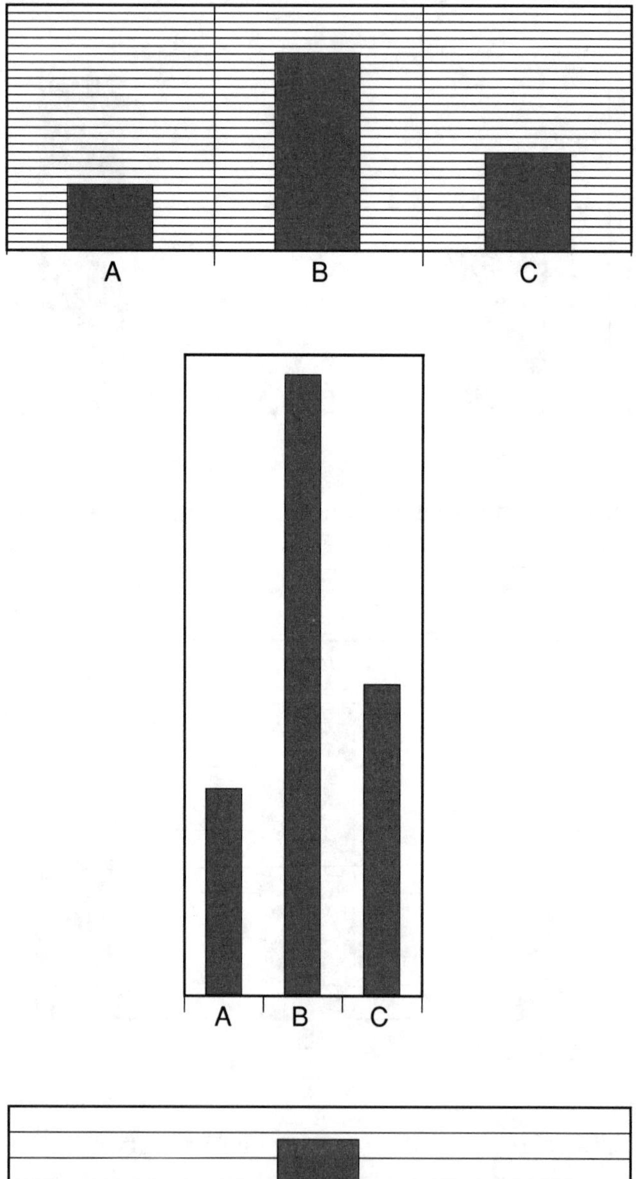

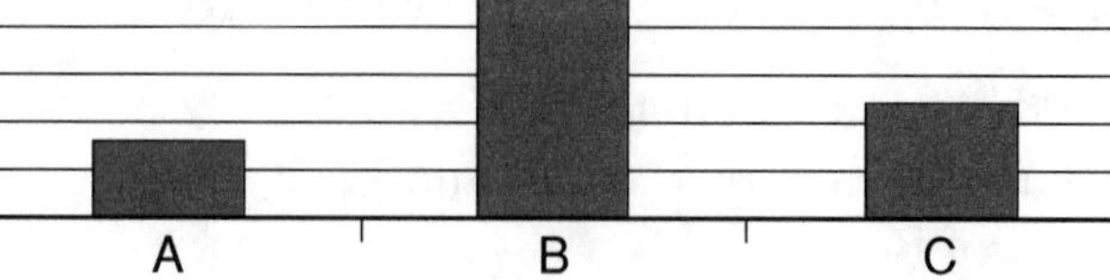

BRAIN TICKLERS
Set # 15

You should find something wrong with each of the following graphs, parts of graphs, or interpretations. Correct the error and explain it.

1.

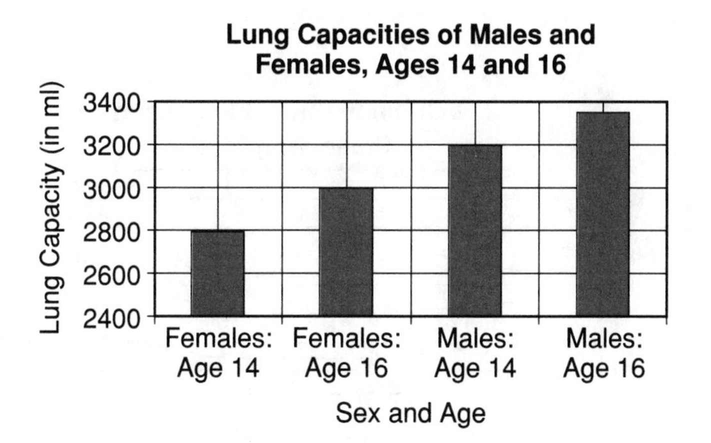

2.

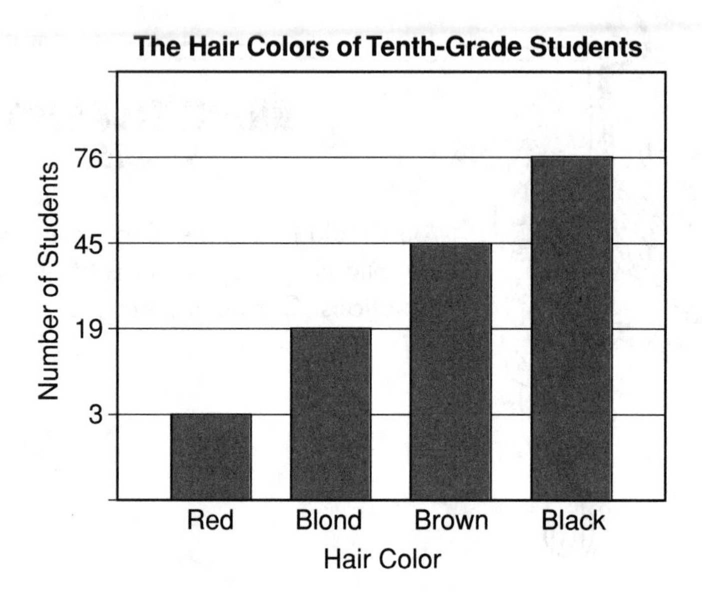
The Hair Colors of Tenth-Grade Students
Number of Students
76
45
19
3
Red
Blond
Brown
Black
Hair Color

3.

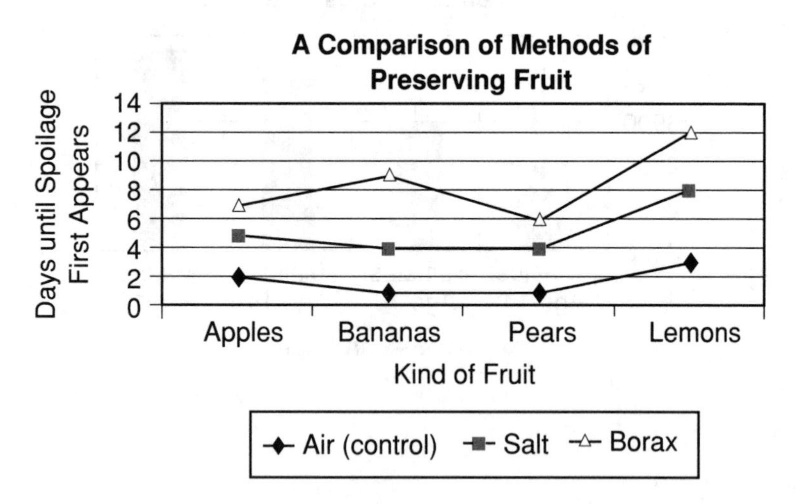
A Comparison of Methods of Preserving Fruit
Days until Spoilage First Appears
14
12
10
8
6
4
2
0
Apples
Bananas
Pears
Lemons
Kind of Fruit
Air (control)
Salt
Borax

4.

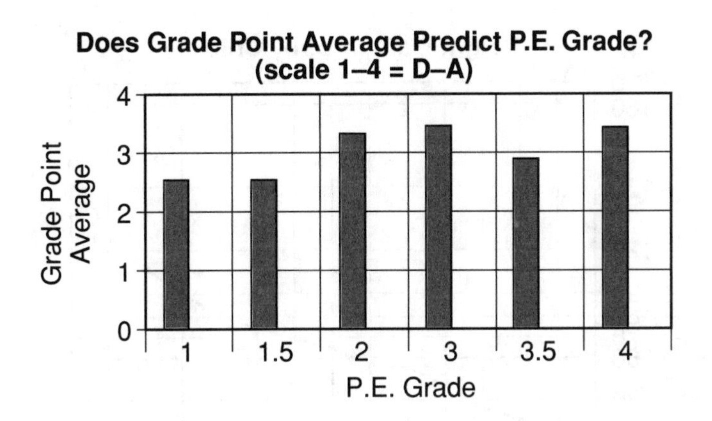
Does Grade Point Average Predict P.E. Grade?
(scale 1–4 = D–A)
Grade Point Average
4
3
2
1
0
1
1.5
2
3
3.5
4
P.E. Grade

5.

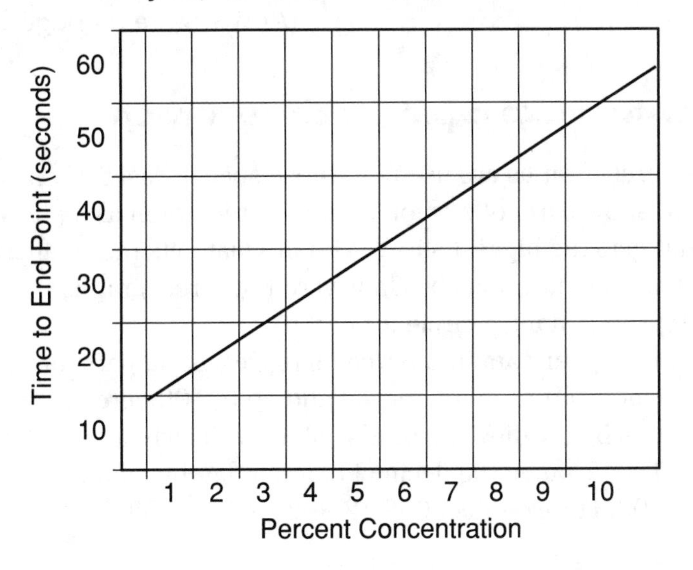
Do Higher Concentrations of Amylase, a Salivary Enzyme, Make the Reaction Go Faster?
Time to End Point (seconds)
60
50
40
30
20
10
1
2
3
4
5
6
7
8
9
10
Percent Concentration

6.

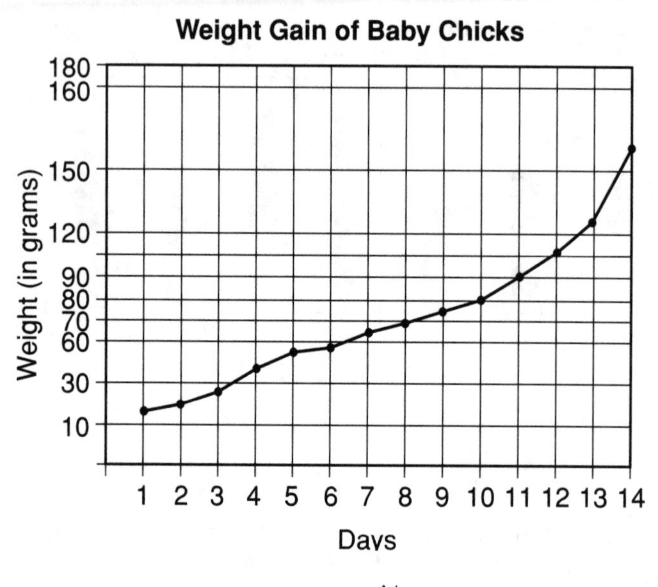

(Answers are on page 205.)

How to make a pie (or circle) graph

Pie or circle graphs are useful in those special cases when you want to show parts of a whole. Think of the graph as a pie. Who or what gets the biggest slice? Who or what gets the smallest? Those are the questions best answered by circle graphs.

Steps in making a circle graph:

1. Turn your data into percents representing parts of the whole. (Your numbers will add up to 100 percent.) Do this by dividing the part's value by the whole, or total. Then change the decimal to a percent by multiplying by 100. For example: $6 + 12 + 29 + 11 = 58$

   - $^{6}/_{58} = 0.10 \times 100 = 10\%$
   - $^{12}/_{58} = 0.21 \times 100 = 21\%$
   - $^{29}/_{58} = 0.5 \times 100 = 50\%$
   - $^{11}/_{58} = 0.19 \times 100 = 19\%$

   Check your work: $10 + 21 + 50 + 19 = 100$ percent (all parts of the whole are accounted for).

Note: Sometimes, because of rounding, you'll find that your total equals 99 percent or 101 percent. It's okay to go back and round one of your numbers up or down until the total comes out to 100 percent.

2. The distance around a circle is measured in degrees. (Not the same as temperature.) A complete circle measures 360 degrees.

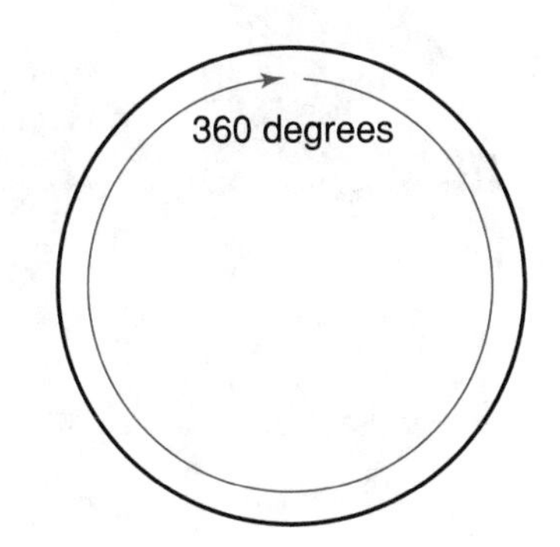

To determine how big each pie slice must be, multiply the percent of each item (expressed as a decimal) by 360.

For example, a 21 percent slice = 0.21×360 degrees
= 76 degrees.

Calculate the size of each pie slice. If your calculations are correct, your degrees will total 360. Once again, you may need to adjust for rounding.

3. Now you will need a compass, ruler, and protractor. Use the compass to draw a circle. Mark the center point. Draw a straight line from the center to any point on the edge. This will be your starting line.

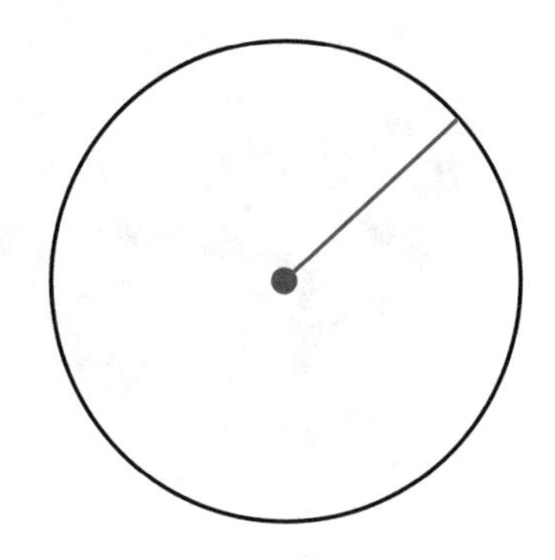

4. Use your protractor to measure angles in degrees from your starting line. (Your teacher can show you how.) Mark off slices of the pie equal to the calculations you have made. For example:

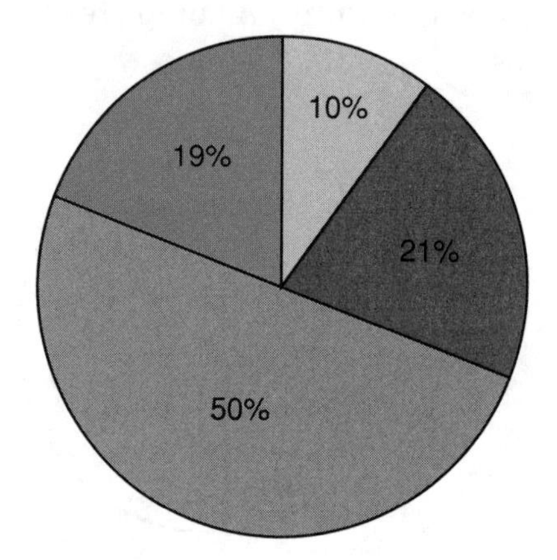

5. Continue marking slices until you include all the data. Title your graph, and label the slices so that anyone who sees it will know exactly what you measured.

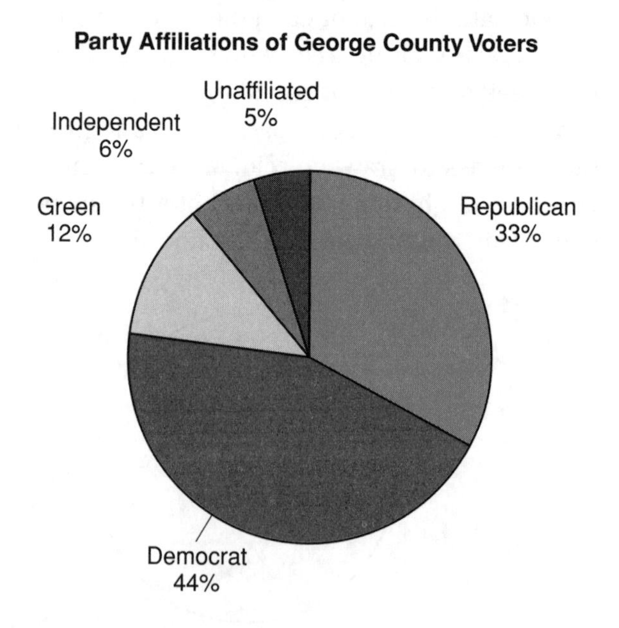

Savvy Science Tip
Some computer graphing software will convert data values into percent automatically when making a circle graph. Check to see what yours will do.

BRAIN TICKLERS
Set # 16

Make a circle graph of the following data:

Flavors of toffee pieces in a kilogram package of mixed flavors: chocolate—34; almond—17; mint—22; butter—53; and raspberry—8.

(Answers are on page 206.)

ADVICE FROM KIDS WHO'VE BEEN THERE, DONE THAT

All these numbers. All these graphs. It seems like so much to learn, so much to understand, so many mountains to climb.

Is it really?

Listen to what these savvy science fair pros have to say:

"I learned a lot just by playing with graphs. I'd make one and see if it told me anything. I ended up trashing a few, but some of them came out terrific."

—Jason Melbourne
Bluefield, WV

"I started statistics with a standard deviation. Three years later, I was explaining regression and analysis of variance to my senior math class. I never would have guessed it. It all seemed so hard at first."

—Medina Oswald
Greensboro, SC

"I made so many tables for my science project, I've started to think in tables. I use them for everything from keeping track of my savings account to making out the cheerleading rosters."

—Josephine Jackson
Seattle, WA

"Tables? Graphs? They're a breeze—once you get the hang of it!"

Juan Giantamo
Orland, CA

Once you've got good tables and graphs, you're about 90 percent of the way to completing that science project.

That's where we're headed in Chapter Five. First, though, here's one more brain tickler to help you tie together everything you have learned.

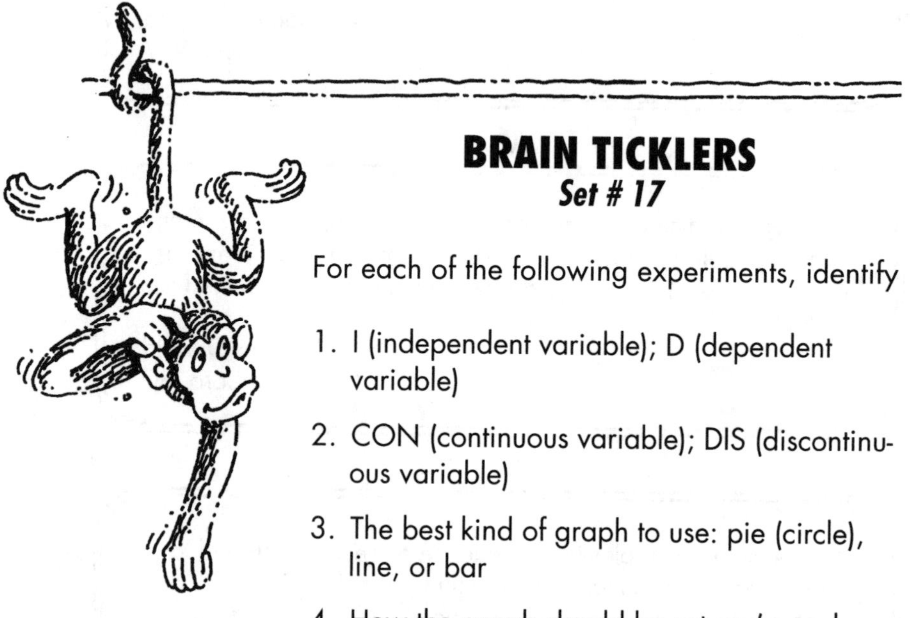

BRAIN TICKLERS
Set # 17

For each of the following experiments, identify

1. I (independent variable); D (dependent variable)
2. CON (continuous variable); DIS (discontinuous variable)
3. The best kind of graph to use: pie (circle), line, or bar
4. How the graph should be set up (*x*- and *y*-axes for line or bar graphs)

This example from Connor's experiment may help get you started:

Material	Mean Dry Weight Before Burying (in milligrams)	Mean Dry Weight After 3 Months (in milligrams)	Mean Change (before minus after)
Plastic	138	124	14
Paper	361	227	134

Material = I and DIS; Weight = D and CON

Best graph: bar

Material = *x*-axis; weight = *y*-axis

1. Brenda tested different materials to see how well they acted as fuses. (Fuses break and stop the flow of electricity when they get too hot.) Her data table looked like this:

	Time Until Breaking			
Material	Trial One	Trial Two	Trial Three	Mean
Steel Wool				
Mylar				
Copper Wire				

2. Marty measured the weight-lifting capacities of levers. He used levers of different lengths and measured the maximum weight lifted.

3. Carlos recorded the number of seconds cats fix their gaze on pictures of different objects.

4. Stephanie measured the amounts of water taken up by plants as a function of the salt concentration of the water.

5. Xavier timed (in seconds) how fast a teaspoon of 10–40 weight motor oil would flow down a pane of glass at different temperatures.

6. Vicki recorded the names and numbers of all birds visiting her feeder between the hours of 3 P.M. and 4 P.M. in September.

(Answers are on page 206.)

BRAIN TICKLERS—THE ANSWERS

Set # 13, page 144

Materials:
1 pot
a handful of small stones
a saucer
a spoon
1 bag of potting soil or soil dug from the garden outside and placed into a bag

1 package of seeds or a seed saved from a flower, fruit, or vegetable
a ruler
water
a measuring cup
a sunny windowsill

1. Place the handful of small stones into the bottom of the pot. This will allow drainage.

2. Set the pot into the saucer. (It will catch any water that may overflow.)

3. Open the bag of soil. Spoon out some soil, and place it onto the small stones in the pot. Keep spooning until the pot is nearly full.

4. Open the package of seeds and pour out one, or get a seed you have saved. Place the seed onto the dirt.

5. Spoon more soil from the bag to cover the seed to a depth no greater than 1 centimeter or the depth recommended on the seed package. (Use the ruler to measure.)

6. Pour $\frac{1}{4}$ cup of water into the measuring cup. Dribble it slowly over the soil. If no water runs out of the bottom of the pot, repeat until the soil is wet all the way through.

7. Set the pot onto a sunny windowsill. Add more water whenever the soil feels dry. Wait for the seed to sprout. Enjoy watching the seed grow.

Set # 14, page 151

Set A.

1. Which material lets through the most (or the least) light? or

How do the light transmission characteristics of some common materials compare?

2. What is the relationship between the mass of an object, the height from which it falls, and the size of the diameter of the crater it makes when it strikes a particular surface? You might imagine possible impact surfaces for this experiment such as fine sand, powder (like the moon or Mars), mud, or some other.

3. Is it true that people break more traffic laws under certain weather conditions? or

   Does the number of traffic tickets given in Meterville correlate with weather conditions?

4. Does color blindness affect one sex more than another? and/or

   Are there more color-blind students in one grade than another in our school?

5. How do the method and length of storage affect the aluminum content of food?

Set B.

6.

Flower	Drying Method	Beginning Weight (in grams)	Weight After Two Weeks (in grams)
Zinnia	Salt		
Zinnia	Borax		
Zinnia	Silica gel		
Marigold	Salt		
Marigold	Borax		
Marigold	Silica gel		

or

Flower	Drying Method (weight change after two weeks, in grams)		
	Salt	Borax	Silica Gel
Zinnia			
Marigold			

7.

Site	Color			Weight		Shape		
	White	Brown	Mixed	<10 Grams	>10 Grams	Round	Spiral	Other
A								
B								

8.

Sample	Weight Before Cooking (in grams)	Weight After Cooking (in grams)	Loss of Weight (before minus after, in grams)
Ground top round	100		
Ground bottom round	100		
Ground chuck	100		
Ground sirloin	100		

9. Time Until Bubbling Stops

Potato Cube (in cm^3)	Hydrogen Peroxide Strength		
	1 Percent	2 Percent	3 Percent
1			
8			
27			

10.

Name	TV Time Daily	Reading Time Daily

And—for data analysis:

Group	Average TV Time	Average Reading Time	Difference (TV minus reading)
Boys			
Girls			
Teens			
Adults			

Set # 15, page 191

1. This graph does not begin with zero on the y-axis. That makes the differences in lung capacity look much greater than they actually are.

2. The numbers of students should not be the points on the number line. Divide the x-axis into equal units from 0 to 50. Data points can then be located along the number line.

3. This should be a bar graph. The lines make no sense. Use line graphs only for continuous variables.

4. The x-axis is laid out improperly. The units aren't equal.

5. Number the lines, not the spaces.

6. This student forgot to plan ahead. Use equal space for equal numbers.

Set # 16, page 197

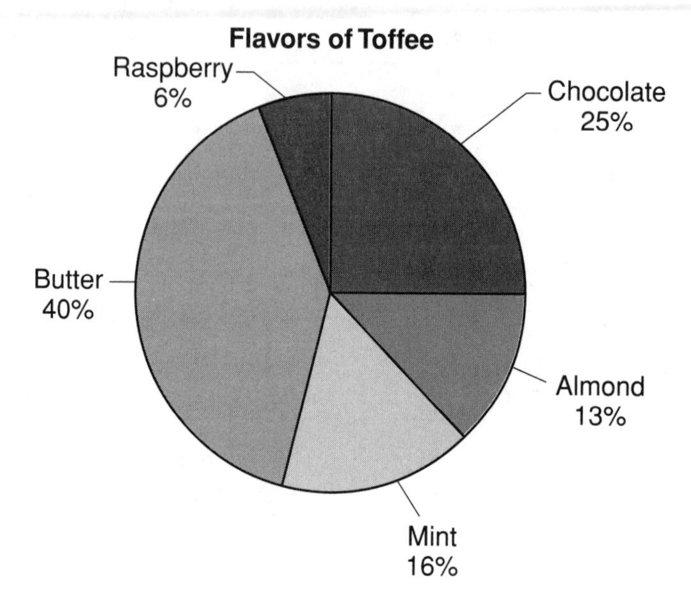

Set # 17, page 200

1. Material = I and DIS; Time = D and CON
   Best graph: bar
   Material = x-axis; time = y-axis

2. Length of lever = I and CON; weight = D and CON
   Best graph: line
   Length of lever = x-axis; weight = y-axis

3. Objects = I and DIS; time = D and CON
   Best graph: bar
   Objects = x-axis; time = y-axis

4. Salt concentration = I and CON; water taken up = D and CON
   Best graph: line
   Salt concentration = x-axis; amount of water taken up = y-axis

5. Temperature = I and CON; flow rate (time) = D and CON
   Best graph: line
   Temperature = x-axis; time (in seconds) = y-axis

6. This is not an experimental project, so variable distinctions do not apply.
   Best graph: circle to show percent of each species observed. A bar graph of numbers and species might also be helpful.

CHAPTER FIVE

Tell the World What You've Found

Who can forget Mr. Spock in the original *Star Trek* TV series and movies? Played so brilliantly by Leonard Nimoy, the Vulcan with the pointed ears lacked emotion—at least most of the time. Logic guided Spock's thinking and often set him at odds with his whimsical, intuitive, and sometimes irrational crewmates.

Science at its best is logical. Scientists aren't Vulcans; they're human beings, so they aren't always logical. Their personal lives are as emotional as the next person's, but when whipping their research into shape and presenting it to others, scientists must use logic.

What is logic? You might call logic a healthy dose of common sense seasoned with plenty of skepticism. Good scientists answer their critics *before* the criticisms come. They spot loopholes in their evidence and fill them with results from additional experiments. They consider all possible hypotheses and test them one by one.

Clear, accurate, unambiguous communication is a must in science. A good scientific report is specific, tight, focused, and detailed enough to explain the what, why, and how of the research. Most important, a good report never draws a conclusion not supported by data.

Now it's your turn to play Mr. Spock. Your data collection is nearly complete. You've written a sound procedure and completed tables and graphs that reveal all. Now what? It's time to tell the world what you've found just as clearly and logically as Mr. Spock would.

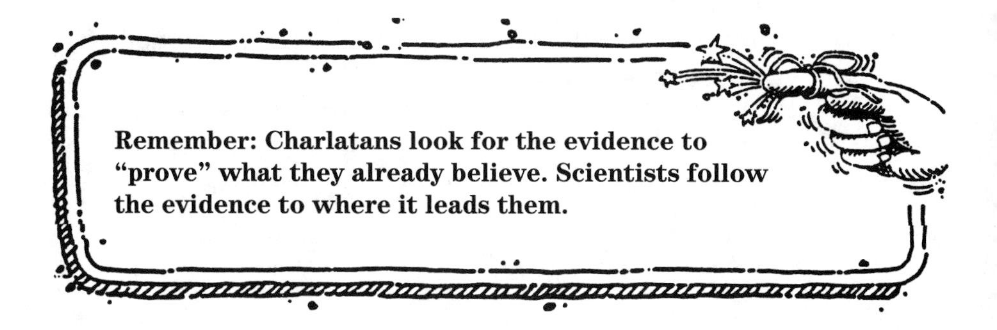

Remember: Charlatans look for the evidence to "prove" what they already believe. Scientists follow the evidence to where it leads them.

YOUR WRITTEN REPORT

Although nonexperimental, experimental, and computational projects require slightly different approaches, all demand a logical sequence of steps arranged in a (more or less) standard way.

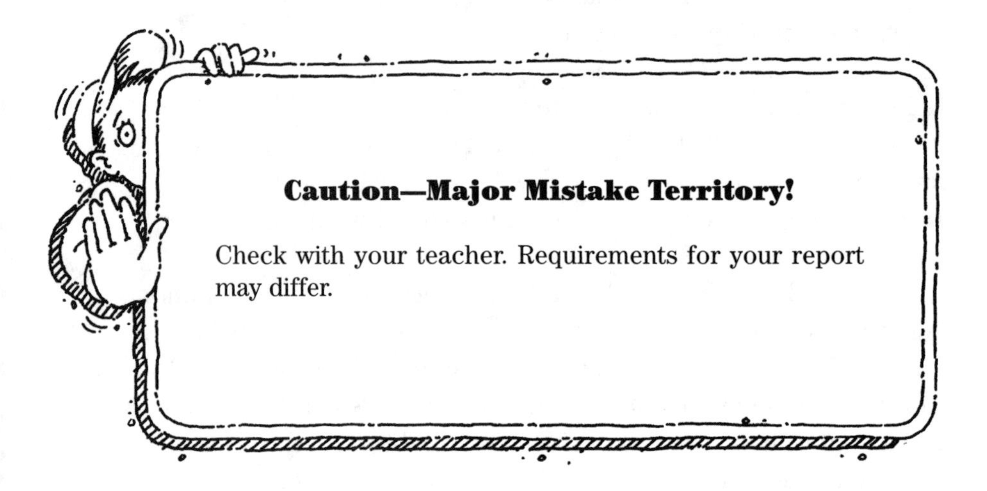

Caution—Major Mistake Territory!

Check with your teacher. Requirements for your report may differ.

Most science project reports should contain the following:

Title page: A specific and descriptive title for your project, your name, and (perhaps) school and date, depending on local requirements

Table of Contents: Titles and page numbers for the major sections of the report: Abstract, Problem(s), Hypothesis, Procedure (sometimes called Materials and Methods), Results, Conclusions (sometimes optional), Discussion (often optional), Sources, and Appendices (if you have attachments you wish to include)

Abstract: A brief summary (no more than one paragraph) of your project and its findings

Problem(s): Your project's purpose(s); the specific question(s) you tried to answer
Hypothesis: Your prediction of what you expected to happen
Procedure or Materials and Methods: Your experiment, study, or construction, explained step-by-step—in chronological, numbered order (review Chapter Four)
Results: Outcomes and observations; descriptions, tables, and graphs showing exactly what you found
Conclusions: Interpretations of your results, explaining whether the hypothesis proved correct or incorrect and why
Discussion (often optional): What you still don't know, what you still need to do, suggestions for new hypotheses and experiments
Sources: Cite the sources of your information and help, whether published articles or personal contact with experts or organizations
Appendices: Attach supporting materials such as copies of articles, correspondence, telephone numbers and addresses, and any other material that supports your project; for computational or computer design projects, put your codes and graphics here

The title page

On your title page, you'll need a title for your project—along with your name and other identifying information as required by your teacher or science fair organizer. Common sense says that your title should describe your work and capture attention, but what exactly makes a good title?

Skim the Table of Contents of a scientific journal, and you'll find titles such as these (which are real!):

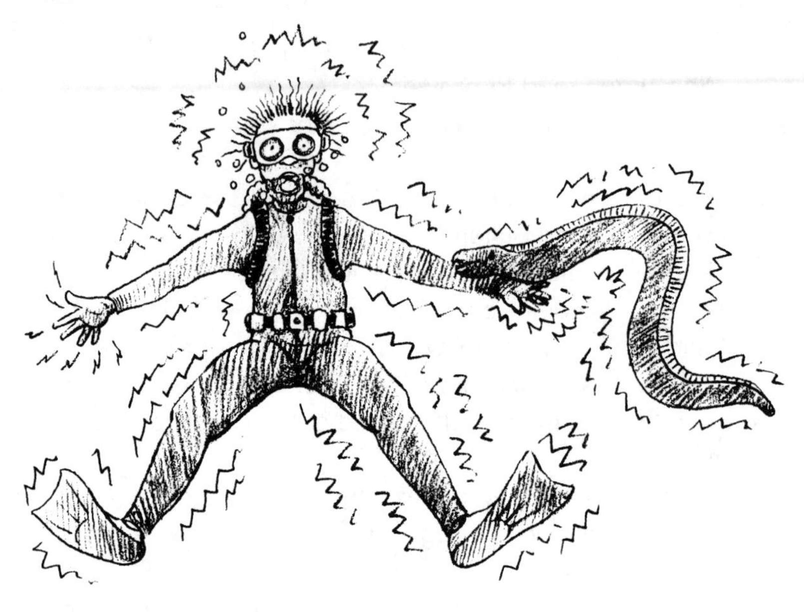

- Neural substrates for species recognition in the time-coding electrosensory pathway of mormyrid electric fish
- Laser frequency-modulated spectroscopy of a laser-guided plasma in sodium vapor: Line positions for NaH (A1Sigma+–X1Sigma+), Na (9–13d and 11–14s), and Ar (5p–4s)
- Detection of multiple stimulus features in a trade-off in the pyramidal cell network of a gymnotiform electric fish's electrosensory lateral line lobe

Why do scientists write such complex titles? They don't love big words and long sentences. They want to tell other scientists exactly what they did and how they did it. Long, detailed titles help other scientists decide quickly whether they can use what's in the report.

You, too, want to convey as much information in your title as possible, but you don't have the luxury of length that scientists enjoy. Because you want your title to attract students, teachers, judges, and visitors to your project, you'll want something shorter (and perhaps snappier) that captures the essence of your project in a few well-chosen words.

The following guidelines and examples may help.

1. Be specific

Poor	Better
Clay	The Clay Content of Titusville Soil
Electricity	Conductivity of Ferrous Metals
Garlic	Does Garlic Retard Bacterial Growth?

2. Keep it short but not too short

Poor	Better
Weather	Rainfall in Meterville: 1990–2000
Eggs	Is One End of an Egg Stronger than the Other?
Amplifiers	Which Materials Amplify Sound Best?

3. Include both independent and dependent variables

Poor	Better
Mice and Mazes	The Effect of Protein Foods on Maze Learning in Mice
Insulation	Ignition Temperatures of Insulating Materials
Parents and Dates	Parents' Attitudes Toward Teen Dating

4. Try asking a question

Poor	Better
Blood Clotting	What Factors Affect Blood-Clotting Time?
Rust Prevention	What Lubricants Prevent Rusting?
X Rays and Safety	What Devices Make X-Raying Safer?

5. Take advantage of subtitles to distinguish the general from the specific

Poor	Better
Recycling Plastic, Paper, and Glass	Recycling: Plastic, Paper, and Glass
The Sun-Protection Factors of Sunscreen	Sunscreens: How Do Their Sun-Protection Factors Compare?
Windmill Designs	Windmills: A Comparison of Three Designs

6. Don't overgeneralize

Poor	Better
Poisons in the Ocean	Are Porpoise Bay Clams Safe to Eat?
Color Blindness in Animals	Do Cats See Colors Better Than Dogs?
Crystallography	Do Crystals Grow Better at Cold Temperatures?

7. Don't scare people away; keep your title simple, informal, and friendly

Poor	Better
Bacterial Decomposition of Biodegradable Materials in Aqueous, Nonsaline Environments	Decay in Freshwater
The Variation in Corrosion of Ferrous Metals as a Function of Acidity and Alkalinity	How pH Affects Rusting

8. Don't mislead, exaggerate, or oversell

Poor	Better
At Last! The Perfect Artificial Leg	Computer-Assisted Designs for Artificial Limbs
An Easy Solution to Probability Problems	A Method of Calculating Independent Probabilities
Harness the Tides for Unlimited Energy	A Design for a Tide-Powered Generator

9. If you can think of a clever play on words, use it

For example, one student called the new paving material he made from waste ashes "Ash-phalt." Another titled her battery-operated system that counts currency for the blind "Money Talks."

Abstract

No, abstract doesn't mean complex or difficult. An abstract is a summary of the entire project. It sums up the problem, hypothesis, results, and conclusions in no more than a paragraph. Although the abstract appears first in your report, write it last. Once you have fully explained all your project's steps and outcomes, boiling it down to a few sentences will come easily.

Sample Abstract

Green plants lose water into the air through the process of transpiration. However, some plants are more efficient at conserving water than others. Such plants are better suited to dry habitats. I compared the transpiration rate of three varieties of geranium by measuring water uptake during periods of light and darkness. I found that transpiration is greatest in the light for all varieties, while the variety Regal showed the highest rate in both light and dark. The variety Patriot showed the lowest transpiration rate in the light, suggesting that it may be suitable for growth in dry environments.

Note that this abstract

- summarizes what is known about the problem;
- explains why the problem is important;
- implies the project's main question, "How do the transpiration rates of three different varieties of geranium compare?"
- specifies the dependent and independent variables, in this case transpiration rate as a function of both variety and light/dark conditions;
- describes the testing procedure—measuring water uptake;
- summarizes the conclusion; and
- speculates on the significance of the findings, that is, the growth of a variety of geraniums in dry areas.

Problem

You can state your problem in a single sentence. It is the question you set out to answer. Make sure you write it as clearly and sharply as you can. If your research grew and you went after several questions, simply list them in this section.

Hypothesis (or hypotheses)

For every problem, suggest a hypothesis. Scientists usually use the *null* hypothesis—that is, no effect will occur. For example, the geranium transpiration experiment might have had two null hypotheses:

- No difference in transpiration rates will occur in the light and the darkness.
- The three varieties will show no differences in transpiration rates.

In student projects, the null hypothesis need not always be used. Check your rules. You may be allowed to predict what you actually believe will happen. In any case, your problem and your hypothesis should be closely related.

Consider these examples:

Problem	Hypothesis
How rapidly do wood frog tadpoles grow in length during the first 30 days after hatching?	Wood frog tadpoles will grow longer at a steady rate—increasing by the same amount each day.
How can suspended solids be removed from water?	My spinning disk machine will remove at least 90 percent of suspended solids from samples of Elk Lake water.

Problem	Hypothesis
How can I build a practical solar oven?	Using my modified solar oven design and construction plans, I'll be able to build a solar oven that works well enough to hard fry an egg in three minutes on a sunny day.
How does a crane lift heavy objects?	Using a model block and tackle, I'll calculate (within an acceptable error range of ±5 percent) the mechanical advantage of the pulley setup a crane uses.
How does the release of treated sewage into Justina Bay affect the microorganisms that live there?	I think I will find fewer kinds and smaller numbers of microorganisms in sewage discharge areas than in unpolluted areas.
Potatoes contain an enzyme that breaks down hydrogen peroxide. What factors affect the speed of that reaction?	I'll time the reaction at different temperatures and pH levels and show that extremes of both variables slow the reaction.

BRAIN TICKLERS
Set # 18

Write two hypotheses for each of these problems. Make the first a null hypothesis. Make the second what you think will happen (if different from the null).

1. Is acceleration greater on the dips or the turns of a roller coaster?
2. Will an increase in carbon dioxide in the air increase plant growth?
3. Does diet affect the concentration of heavy metals in hair?
4. Does pupil diameter change in darkness and light?
5. Do right-handed people have a wider range of peripheral vision than left-handed people?
6. How does salt concentration affect the density of water?
7. How does the salt concentration of water affect conduction of electricity?
8. Which of three brands of detergent removes stains the best?
9. How does the water content of different dried fruits vary?
10. What is the pH of tap water in the typical home in Meterville?

(Answers are on page 242.)

Procedure (or materials and methods)

As you have already learned, your procedure must

- list all materials, equipment, and supplies needed; leave out nothing
- describe each step from beginning to end—in order—again, leaving out nothing

As you collected data, you probably improved on your procedure. You need not throw away data from early trials and unimproved procedures. You may wish to report in this section how your procedure evolved, opening the way for refining your results and conclusions as detailed in later sections. This may be especially important for design-redesign and computational projects where the procedure *is* the goal.

This is also the spot for photographs and drawings showing how you did things. Save visual records of outcomes for the next section.

Results

Here's the place for all those statistical tests, tables, and graphs you completed while analyzing your data (see Chapter Four). Include only your best, taking care that everything you include is both complete and accurate in every respect. (See "Checklist for Tables and Graphs" in Chapter Six.)

Write a paragraph or two to accompany each statistical test, data table, and related graph(s). Explain exactly what your numbers mean. Point out comparisons or significant trends as you discuss each graph. For example:

- *A* scored consistently higher than *B* on all measures.
- The trend noted during the initial growth phase failed to persist into the second and third weeks.
- Results were mixed and depended on environmental temperatures.
- The second trial of the experiment failed to confirm results from the first.

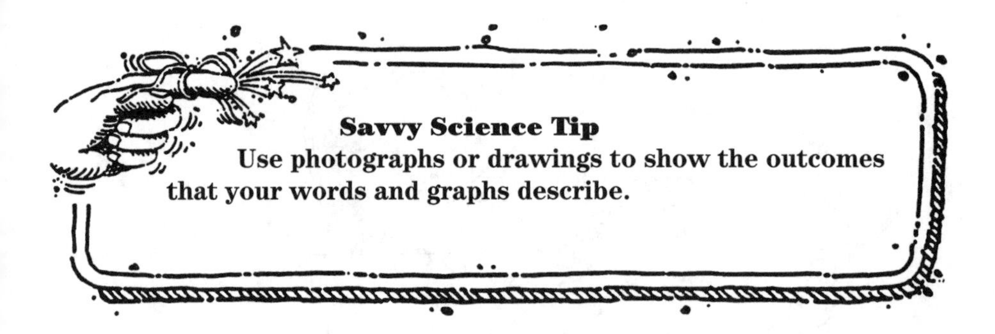

Savvy Science Tip

Use photographs or drawings to show the outcomes that your words and graphs describe.

Conclusions

Here's the place to comment on the importance of your data. When writing your conclusion, go back to your hypothesis and discuss whether your original prediction proved true or false. Here are some conclusions based on the previous hypotheses:

Hypothesis	Conclusion
Wood frog tadpoles will grow longer at a steady rate—increasing by the same amount each day.	My hypothesis was incorrect. The tadpoles grew most rapidly during the first 17 days. Growth slowed on days 18–22 and stopped on day 23, when the tadpoles attained maximum length.
My spinning disk machine will remove at least 90 percent of suspended solids from samples of Elk Lake water.	Using a spectrophotometer borrowed from a university lab, I was able to measure an average 75 percent reduction in suspended solids in my water samples.
Using my modified solar oven design and construction plans, I'll be able to build a solar oven that works well enough to hard fry an egg in three minutes on a sunny day.	I exceeded my expectations. My oven could hard fry an egg in two minutes even on a cloudy day.
Using a model block and tackle, I'll calculate (within an acceptable error range of ±5 percent) the mechanical advantage of the pulley setup a crane uses.	Drawing the rope 4 decimeters raises the load approximately one decimeter. The mechanical advantage of a crane's block and tackle is approximately 4.
I think I will find fewer kinds and smaller numbers of microorganisms in sewage discharge areas than in unpolluted areas.	I found fewer kinds of organisms but in larger numbers, so my hypothesis was partly correct and partly incorrect.
I'll time the reaction at different temperatures and pH levels and show that extremes of both variables slow the reaction.	The reaction is slow at high and low pHs and at high and low temperatures. My hypothesis was confirmed.

BRAIN TICKLERS
Set # 19

Reconsider each of the problem statements given in Brain Ticklers Set #18 on page 222. Assume your results showed a positive relationship between variables. Write an appropriate conclusion statement. (Hint: You can write a more specific conclusion if you make up some numbers. Who knows? You might actually be able to conduct the experiment to see if you guessed the correct numbers.)

(Compare your guesses with those on page 243.)

Discussion

Although scientists usually conclude their research reports with a discussion section, students often do not, and some science fair rules omit this section. If it is permitted or required for your project, include in this section

- guesses about why something in your project may have gone wrong;
- new questions that your research brought to light;
- ideas for how to do better next time;
- hypotheses for new experiments to conduct in the future; and
- hopes for the application of your research (perhaps treating disease or improving the quality of the environment).

If you want to speculate, here's the place to do it. Here, also, is the spot for planning how your project will continue next year and the year after that.

Sources

List all articles, book, pamphlets, sound recordings, videos, and any other published material from which you obtained information. (See Chapter Two for the forms citations can take.) Also list the people who helped you and the help they provided.

Appendices

Attach to your report any additional material that will help an interested reader understand your project. A particularly relevant article may become an appendix. Appendices can include sources of materials and supplies, letters received from experts, logs of telephone conversations, code for computer projects, designs for devices or instruments, and so on. Some science fairs also require an accounting of the money spent on a project. That can go in the appendices, too. Designate each appendix by a letter, and list it in your Table of Contents. For example:

APPENDICES

Appendix A: Letter and Article from Dr. Juan Geraldo, MIT
Appendix B: Decay Rates of Radioactive Isotopes
Appendix C: Sources of Materials, Equipment, and Supplies
Appendix D: Project Budget and Receipts

Appendices do not need page numbers.

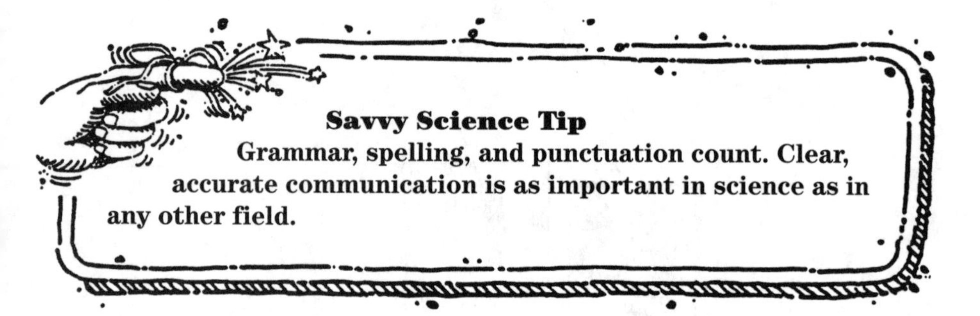

When Disaster Strikes

The worst has happened. Your experiment has failed. Your seeds didn't sprout. Your crystals crumbled. Your electric car spins in

circles. Whatever you thought you could prove, you can't. In fact, your results show that you've been on the road to ruin from the beginning. Is everything lost?

No way! Working scientists often find themselves in exactly the same position. The trick is to turn *negative data* to positive advantage. Take a cue from the pros and turn bad news into glad tidings with the **three *Ts*: Tell. Think. Test.**

STEP 1. TELL

Above all, science values truth. If you planted 47 radish seeds and 46 of them didn't come up, say so. Make graphs of your numbers no matter how silly they seem. Display your empty seed trays even if they look pitiful.

STEP 2. THINK

What factors might explain your results, dismal as they were? For example, if your seeds didn't sprout, can you think of reasons why? Did you put them into a cold place? Did you overwater or underwater? Did you plant them too deep or in the wrong kind of soil? Think about your explanations not as failures but as fresh hypotheses.

STEP 3. TEST

Plan and conduct some experiments to test one or more of your new hypotheses. You may not have much time left, but a simple experiment may help unravel part of the mystery. For example, some radish seeds sprinkled on wet, damp, and dry sponges can quickly reveal (within two or three days) how water affects the germination (sprouting) process.

If your experiment went wrong, use your Results section to say so. It's all right to report that

- your data failed to support your hypothesis;
- test results were mixed;
- measures of error (such as standard deviation) showed unacceptably large values; or
- results from some trials contradicted others.

If you were able to isolate and control a source of error, explain how your thinking changed and what you did about it.

There's no shame in failure. However it's a shame to quit just when you've started to learn so much.

YOUR DISPLAY

Your display should include almost all the same parts as your written report. At a minimum, it should include the

- problem,
- hypothesis,
- procedure,

- results, and
- (perhaps) conclusion.

Your display will not give as much detail as your report, however. Design your display to grab the viewer's attention and get your main points across clearly and quickly. In many cases, you won't be present to explain your project. Your display must do your explaining for you.

Most science fairs require a backboard of some kind, often with three sides. Science fair rules get very specific about the size and shape of the backboard. Ignore them, and you may find yourself disqualified. Students sometimes construct their backboards from poster board, but self-supporting materials such as corkboard, Peg Board, or paperboard are sturdier and more attractive.

Some attention grabbers

1. *Make it big.* Find out the maximum allowable size for your display and use it. Less is more only if miniaturization is the key to your project. Apply large lettering that people can easily read from a distance—especially for the title and main headings. Do it by hand, use press-on

letters from an art supply store, or have your computer print a hard copy in big type that you then paste onto your display board.

2. *Use bright colors*. Colored paper can frame blocks of type. Color photographs and multicolored graphs will attract attention. Use the biggest, best, and brightest drawings, photographs, data tables, and graphs you can possibly make. Of course, they must be perfectly accurate. They should also be neat, neat, neat.
3. *Keep your display clean*. Wrap it in plastic whenever you move it.
4. *Display objects whenever possible*. On the table in front of your backboard, you have room to display something associated with your project. If you grew plants, your experimental subjects can tell their own story. If possible, show something that invites observers to collect data of their own. For example, a thermometer in a model greenhouse lets visitors read the temperature for themselves.
5. *Display a copy of your written report and your logbook*. Your logbook is a compilation of the best things from your journal. Use it to support your findings and provide additional detail for judges, teachers, and visitors who want to know more.

> International Science and Engineering Fair rules prohibit the display of a wide variety of items. See "Do Not Display," Chapter Six, for a partial list. Always check with your teacher before deciding what you can display.

A dozen definite don'ts for your display

Don't

1. use fuzzy photocopies or unfocused pictures
2. copy information from encyclopedias, books, magazines, or other sources
3. scribble, cross out, smudge, or otherwise deface your display
4. write or draw in pencil
5. paste lined notebook paper onto your display
6. use tiny letters or fancy, hard-to-read scripts
7. squeeze too much onto your display
8. include too little information in your display
9. use words without pictures
10. use words and pictures without numbers
11. use numbers without making a graph to go with them
12. make it pretty without making it meaningful

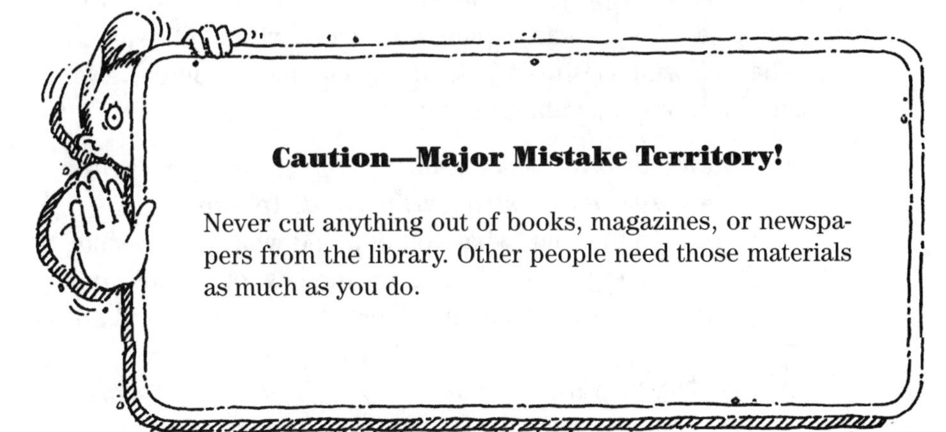

Caution—Major Mistake Territory!

Never cut anything out of books, magazines, or newspapers from the library. Other people need those materials as much as you do.

YOUR ORAL PRESENTATION

Caution—Major Mistake Territory!

Check the rules, and know what's expected of you. Some fairs give you a chance to be personally interviewed. You can explain your work and show the judges how much you've learned. Other fairs keep you away from your display. Your display and written report must speak for you.

In some science fairs, you'll be asked to stay with your project and answer questions from judges and visitors. Don't panic. If you've taken your project seriously, you know more about it than anyone else, and you can explain your work with confidence.

The following tips may make the questions and interviews easier and more enjoyable for you.

1. *When asked to explain your project, keep your answers simple and straightforward.* Tell what you wanted to find out, what you did, and what your results showed. Judges will like to hear the new questions your project has stimulated and where your research will go next.
2. *Practice explaining your project in your own words.* Test your presentation on family and friends, but don't memorize a "canned" speech. You'll sound awkward, nervous, and unsure of yourself.
3. *Anticipate questions.* If your plants died, be ready to suggest a reason why. If your results were mixed and hard to interpret, say so and offer a range of alternative explanations. If your data failed to support your hypothesis, prepare to explain why that outcome is reasonable and valuable.

4. *Look and feel your best*. Take care of yourself. Get enough sleep. Eat a healthy breakfast. Wear clean, comfortable, attractive clothes that make you feel good about yourself.

5. *Defeat shyness*. Stand straight. Square your shoulders. Don't talk to the floor; talk to the judge. In your mind's eye, see yourself as the picture of confidence. Act confident (even when you don't feel it) and soon—amazingly—you'll start to feel confident. Try it. It really works!
6. *Don't get defensive*. Questions aren't criticisms or attacks. The visitor or judge who asks you a question is sincerely interested in your answer. Don't assume questions are intended to trip you up or reveal your ignorance.

Handy Hint

When someone asks a question you can't answer, how should you respond? A simple "I don't know, but I think I could find out," works every time. If you're quick enough to suggest a specific way to find out, you'll come off like a star.

7. *Judges and teachers are people just like you.* Maybe your science fair judges look scary and have long strings of titles and degrees after their names. Don't worry. They're just people, and they're people who are interested enough in you and your work to take the time to come to your science fair. Offer them a smile and your best effort, and you'll surely get the same in return.
8. *Remember what's important.* Ten years from now, you probably won't worry much about whether you won or lost the science fair. However, you will remember what you did, what you learned, and how happy you were with your effort. If you satisfy yourself with your best possible performance, you'll satisfy teachers and judges as well.

Athletes may not win every time, but they consistently strive for their personal best. Are science students any different?

WHAT TEACHERS AND JUDGES LOOK FOR

The big day arrives. You've finished your research. You've completed your logbook. Your written report is organized, accurate, and neat. Your display sparkles. You're wearing your favorite sweater. What next? If you have completed a project for a science fair, you may now face the judging process. Experts and volunteers will look at your work and, perhaps, talk with you about it. You'll probably feel nervous. That's only natural. But did you ever think that judges feel nervous too? Here's one judge's story, in her own words.

Science Fair...or Science Unfair? One Judge's Perspective*

by Barbara Foster

I hesitated outside the gymnasium door. Along with other judges, I had been briefed on the science fair's rules. I had been given enough evaluation forms to rate the 11 in my assigned category—physical science. The time had come to view the projects and listen to the students explain them. And I was nervous.

I wondered: Will I judge fairly a project on a topic I know nothing about, compared to how I will judge a project on a topic I'm familiar with? Will I be able to see the merit in a project done on a subject matter that doesn't interest me? In other words, will I help make this a *fair* science fair, or an *unfair* science fair? I breathed deeply and entered the gym-turned-judging area.

For an hour I listened to students describe how potato batteries work, discuss whether earthen dams or hydroelectric dams are better, explain how pistons work in go-cart engines, and answer a question we've all pondered: Why is the sky blue? As I listened to each student, I considered: Was the procedure appropriate for studying the stated hypothesis? Is the conclusion logical? And I looked around: Are the steps involved in investigating the hypothesis clearly outlined? Is there a record of the data collected?

And soon I found my fear that I might judge according to my own biases fading away. If a project was exceptional, it stood out. If the student had done a poor job, it was obvious. One student had hypothesized that a snowboarder making sharp turns going down a hill would cover a given distance faster than the same snowboarder making wide sweeping turns. But he had made only two trials, one with each type of turn—not enough to address his hypothesis adequately. Another student was more thorough. In the middle of her experiments her hygrometer (an instrument used to measure atmospheric humidity) had malfunctioned. She explained how she discovered and corrected the problem, and showed records not only of the original flawed data but also of the data obtained from her retrials.

When one student had to read off his poster to answer my questions, I knew he didn't know his own results. It didn't matter that I wasn't familiar with the topic (the dynamics of airplane flight), because it was clear that he wasn't either. By contrast, another entrant could explain, without looking in his logbook, any point covered in his 20 pages of notes. I liked the topic of that entry—how solar-powered cars work. However, I gave that project the highest rating not because of my own interest in the subject but because the student, with his volumes of research and his homemade apparatus that demonstrated the project's concept, earned it.

So, even though I went to the judging with certain biases, I handed him my evaluation forms feeling satisfied that I had, indeed, helped to make the competition a science *fair*.

*From ODYSSEY's November 1997 issue: *Science Olympics—Ready, Set, Win!*, © 1997, Cobblestone Publishing Company, 30 Grove Street, Suite C, Peterborough, NH 03458. Reprinted by permission of the publisher.

This judge's story emphasizes the four criteria most judges—and most science fairs—apply when assessing projects. Think about them as you work to make your project as good as it can be!

1. *Knowledge achieved.* More than anything else, teachers and judges want to know that you've learned a lot from your work. Become an expert about your topic. Your learning and enthusiasm will then show in your report, your display, your logbook, and your oral presentation.
2. *Effective use of the scientific method.* Here's where logic comes in. Have you collected enough data? Have you interpreted it carefully and skeptically, not overstating your conclusions? Have you tested and retested? Have

you made your tables and graphs accurately? Have you described their meaning correctly?

3. *Clarity of expression.* Your meaning should be clear, even to judges and visitors who know nothing about your topic. Here's where a neat, attractive display that contains no spelling or grammar mistakes comes in.
4. *Originality and creativity.* Is your project something special that stands out from the crowd? Have you explored an idea that is different from everyone else's? Have you presented your work in an engaging and attractive manner?

WHAT NEXT?

Have you ever experienced the postvacation blues? You come home from the trip of a lifetime only to feel let down—as if your ordinary life were a bit more dull and boring than you ever realized. The days and weeks after a science fair can feel a little like the postvacation blues, especially if you've put your all into your work. The buildup to the event and the excitement of the fair followed by a return to everyday life can make you feel a little sad.

Fight the post–science fair blues by continuing your enthusiasm for science in general and your project in particular. If you're planning to complete a science project next year, get a head start. Many projects that win big prizes take several years to complete. Ask new questions, state new hypotheses, collect fresh data. Start building that piece of equipment you didn't quite get to this year. Write that letter to an expert that you planned but never finished. Most important, develop your interests and abilities in science. Go to science museums. Watch science shows on television. Read science books that interest you. Keep that questioning attitude.

The world is your laboratory. Watch and wonder. Life is the grandest science project of all. Live it to the fullest. By the time you've completed your first science project, you shouldn't be afraid to dive into another. Gone is that fear of the great sea of science.

"Come on in. The water's fine!"

Ask courageous questions. Do not be satisfied
with superficial answers. Be open to wonder
and at the same time subject all claims to logic,
without exception, to intense skeptical scrutiny.
Be aware of human fallibility.
Cherish your species and your planet.

—Carl Sagan (1934–1996)

BRAIN TICKLERS—THE ANSWERS

Set # 18, page 222

1. *Null:* No difference in acceleration will occur on the dips of a roller coaster as compared with the turns.

   Predictive: I think greater acceleration will occur in the dips of the roller coaster than on the turns.

2. *Null:* An increase in atmospheric carbon dioxide will have no effect on the rate of plant growth.

   Predictive: An increase in atmospheric carbon dioxide will increase the rate of plant growth.

3. *Null:* No relationship will exist between diet and the heavy metal content of hair.

   Predictive: Those who consume a diet rich in heavy metals will have more heavy metals in their hair than those who eat a diet low in heavy metals.

4. *Null:* Pupil diameter will be the same in light and darkness.

   Predictive: On the average, pupil diameters will be wider in the dark than in the light.

5. *Null:* Right-handed and left-handed people will not differ in the range of their peripheral vision.

   Predictive: Right-handed people will have a wider range of peripheral vision than left-handed people.

6. *Null:* Salt concentration will have no effect on the density of water.

   Predictive: The higher the concentration of salt in water, the greater its density will be.

7. *Null:* Salt concentration will have no effect on the conduction of electricity in water.

   Predictive: An increase in salt concentration will increase water's conductivity of electricity.

8. *Null:* No observed difference will occur in the stain-removal capabilities of the three products.

   Predictive: I think Stain-Off will remove chocolate, coffee, and blood stains best.

9. *Null:* Different dried foods will have no difference in water content.

   Predictive: I think prunes will have more water per gram than dried peaches or apricots.

10. *Null:* The pH of Meterville water will be neutral, 7.0, on the pH scale.

    Predictive: The pH levels of Meterville tap water will vary between 6.0 and 8.0 on the pH scale, with an average around 7.0.

Set # 19, page 226

1. Acceleration proved to be 37 percent greater on the dips of the roller coaster than on the turns.

2. An increase of 25 percent in atmospheric concentration of carbon dioxide produced a tripling of plant growth rate.

3. Those consuming a diet rich in heavy metals displayed an average 17 percent greater concentration of heavy metals in their hair.

4. On the average, pupil diameter increased by 12 percent in the dark as compared with the light.

5. Right-handed people exhibit a 15° wider range of peripheral vision than left-handed people.

6. Water density increased proportionately with increasing salt concentration.

7. My graph shows that the higher the salt concentration, the greater the conductivity of water.

8. Handi-scrub removed chocolate and blood stains better than its two competitors, but Stain-Off did the best job on coffee.

9. The water content of dried prunes proved 29 percent greater than that of peaches and 16 percent greater than apricots, by weight.

10. The pH values of tap water in Meterville homes ranged from a low of 6.3 to a high of 7.2, with a mean of 6.7. Meterville's water is slightly acidic. The source of the acidity is unknown.

CHAPTER SIX

A Dozen Dandy Checklists to Move Your Project Along

DO-IT-YOURSELF PLAN AND SCHEDULE

Photocopy this page and paste it into your journal. Use it to keep track of your project. Your teacher may set deadlines for you, or you may set your own.

Note: The order of some steps may vary depending on how you work and the nature of your project.

Note: Some steps may require hours. Some may require weeks. Plan accordingly.

Name__________________ Beginning date_____ Date due______

Question____

Step	Deadline	Completed
1. Begin journal	________	________
2. Decide on topic	________	________
3. Select project type	________	________
4. Narrow to a specific question	________	________
5. Request human/animal permission if needed	________	________
6. State hypothesis	________	________
7. Conduct library/on-line research	________	________
8. Cite sources in proper form	________	________
9. Contact experts	________	________
10. Request appropriate help	________	________
11. Prepare project budget	________	________
12. Locate materials, supplies, and equipment	________	________
13. Pin down variables and controls	________	________
14. Learn measuring techniques/devices	________	________
15. Draft interview/questionnaire (if using)	________	________
16. Determine sampling technique (if appropriate)	________	________
17. Recruit subjects (if needed)	________	________

18. Write procedure ________ ________
19. Plan safety procedures/collect safety equipment ________ ________
20. Test procedure ________ ________
21. Prepare tables for data collection ________ ________
22. Collect data ________ ________
23. Perform basic mathematical analysis ________ ________
24. Perform advanced statistical analysis ________ ________
25. Make graphs ________ ________
26. Copy from journal to logbook ________ ________
27. Write report ________ ________
28. Build display ________ ________
29. Prepare oral presentation ________ ________
30. Science fair/project day ________ ________
31. What next? ________ ________

WHAT AM I LEARNING? WATCH YOUR KNOWLEDGE AND SKILLS GROW

Make a copy of this page and the next. Paste them into your journal. Every Friday morning, make a check beside those phrases that best describe your learning experiences during the previous week.

	Week Number														
This week I	**1**	**2**	**3**	**4**	**5**	**6**	**7**	**8**	**9**	**10**	**11**	**12**	**13**	**14**	**15**
Asked a lot of questions	☐	☐	☐	☐	☐	☐	☐	☐	☐	☐	☐	☐	☐	☐	☐
Had some good ideas	☐	☐	☐	☐	☐	☐	☐	☐	☐	☐	☐	☐	☐	☐	☐
Cleared up some confusion	☐	☐	☐	☐	☐	☐	☐	☐	☐	☐	☐	☐	☐	☐	☐
Set a goal	☐	☐	☐	☐	☐	☐	☐	☐	☐	☐	☐	☐	☐	☐	☐
Solved a problem	☐	☐	☐	☐	☐	☐	☐	☐	☐	☐	☐	☐	☐	☐	☐
Learned some new facts	☐	☐	☐	☐	☐	☐	☐	☐	☐	☐	☐	☐	☐	☐	☐
Found a way to do something new	☐	☐	☐	☐	☐	☐	☐	☐	☐	☐	☐	☐	☐	☐	☐
Wrote a procedure	☐	☐	☐	☐	☐	☐	☐	☐	☐	☐	☐	☐	☐	☐	☐
Asked someone for help	☐	☐	☐	☐	☐	☐	☐	☐	☐	☐	☐	☐	☐	☐	☐
Met or corresponded with an expert	☐	☐	☐	☐	☐	☐	☐	☐	☐	☐	☐	☐	☐	☐	☐
Corrected a mistake	☐	☐	☐	☐	☐	☐	☐	☐	☐	☐	☐	☐	☐	☐	☐
Distinguished a fact from an opinion	☐	☐	☐	☐	☐	☐	☐	☐	☐	☐	☐	☐	☐	☐	☐
Recognized a bias	☐	☐	☐	☐	☐	☐	☐	☐	☐	☐	☐	☐	☐	☐	☐
Learned a new technique	☐	☐	☐	☐	☐	☐	☐	☐	☐	☐	☐	☐	☐	☐	☐

	Week Number														
This week I	1	2	3	4	5	6	7	8	9	10	11	12	13	14	15
Kept my journal	☐	☐	☐	☐	☐	☐	☐	☐	☐	☐	☐	☐	☐	☐	☐
Wrote something pretty good	☐	☐	☐	☐	☐	☐	☐	☐	☐	☐	☐	☐	☐	☐	☐
Designed or completed a data table	☐	☐	☐	☐	☐	☐	☐	☐	☐	☐	☐	☐	☐	☐	☐
Designed or completed a graph	☐	☐	☐	☐	☐	☐	☐	☐	☐	☐	☐	☐	☐	☐	☐
Mastered some new math skill	☐	☐	☐	☐	☐	☐	☐	☐	☐	☐	☐	☐	☐	☐	☐
Learned a new statistical technique	☐	☐	☐	☐	☐	☐	☐	☐	☐	☐	☐	☐	☐	☐	☐
Worked on my logbook	☐	☐	☐	☐	☐	☐	☐	☐	☐	☐	☐	☐	☐	☐	☐
Worked on my written report	☐	☐	☐	☐	☐	☐	☐	☐	☐	☐	☐	☐	☐	☐	☐
Worked on my display	☐	☐	☐	☐	☐	☐	☐	☐	☐	☐	☐	☐	☐	☐	☐
Worked on my oral presentation	☐	☐	☐	☐	☐	☐	☐	☐	☐	☐	☐	☐	☐	☐	☐
Felt more confident about my project	☐	☐	☐	☐	☐	☐	☐	☐	☐	☐	☐	☐	☐	☐	☐
Surprised myself with how well I did	☐	☐	☐	☐	☐	☐	☐	☐	☐	☐	☐	☐	☐	☐	☐
Saw a way to do more	☐	☐	☐	☐	☐	☐	☐	☐	☐	☐	☐	☐	☐	☐	☐
Saw a better way to do things	☐	☐	☐	☐	☐	☐	☐	☐	☐	☐	☐	☐	☐	☐	☐
Answered a question I didn't know I could	☐	☐	☐	☐	☐	☐	☐	☐	☐	☐	☐	☐	☐	☐	☐
Didn't give up	☐	☐	☐	☐	☐	☐	☐	☐	☐	☐	☐	☐	☐	☐	☐
Felt successful—a little or a lot!	☐	☐	☐	☐	☐	☐	☐	☐	☐	☐	☐	☐	☐	☐	☐

SAFETY FIRST

Caution: No checklist of safety tips is ever complete. Safety depends most on awareness and vigilance. **Think safety**. Look at what you are doing and recognize hazards *before* any harm is done. This list may help, but your Number One Safety Tool is *you*!

Work habits

- ☐ Disorganization invites accidents. Gather all your materials ahead of time in a clear, clean space big enough for your experiment. Make it neat and organized. Know what you are doing *before* you begin.
- ☐ Ask for help from a responsible person. Better safe than sorry.
- ☐ Don't try to lift objects that are too heavy for you.
- ☐ Never work alone at home or in the laboratory. Make sure a responsible person is nearby to assist if you have an accident.
- ☐ Learn the emergency telephone number for your community. (In most places, it's 911, but check *before* you need it.)
- ☐ Take classes in first aid so you can help others in case of an emergency.

Living things

- ☐ Never collect animals from the wild. Stay away from snakes, turtles, insects, squirrels, chipmunks, mice, rats, rabbits, ticks, mites, bees, and other organisms that are poisonous or can carry disease.
- ☐ Never collect plants from the wild. Some are poisonous. Some are endangered species.
- ☐ Never collect plants or animals from public or private land without permission.
- ☐ Don't taste unknown plants, seeds, berries, or fruits.
- ☐ Wash your hands before and after working with plants.

- ☐ Purchase seeds, fish, plants, or animals for your project from a reputable pet shop or aquarium supply store. Purchase seeds and plants from a reputable garden supply shop.
- ☐ Make sure animals have clean, well-ventilated, spacious, safe living quarters and clean food and water at all times.
- ☐ As a general rule, don't use chemical pesticides or herbicides. If you must use them, seek expert help.
- ☐ Handle animals—even pets—with care. Many bite, scratch, or kick.
- ☐ Wear gloves when handling animals. Wash your hands before and after any contact.
- ☐ Never tease an animal.
- ☐ Never touch an animal you don't know.
- ☐ If you go away, arrange for someone to care for your animals or plants.

Electrical devices

- ☐ Wait until electrical devices cool before touching them.
- ☐ Grasp the plug, not the cord, when removing the plug from the wall socket.
- ☐ Don't use extension cords.
- ☐ Don't use electrical devices near water.
- ☐ Make sure your hands are dry before you touch any electrical equipment.
- ☐ Never connect the positive terminal of a battery to its negative terminal. The wire can get hot enough to cause fires or burns.
- ☐ Use three-prong (grounded) plugs on all tools and electrical equipment.
- ☐ Keep your hands and feet away from moving parts and heating elements.
- ☐ Wear goggles when operating electrical or mechanical equipment.
- ☐ Wear ear plugs when operating noisy machinery.
- ☐ Use care in every situation. Even everyday devices—such as fish tank heaters, motors, soldering irons, hot plates, pencil sharpeners, and electric fans—can be dangerous.

Glass

- ☐ Whenever possible, use plastic, not glass.
- ☐ Prevent cuts. Grind or sand sharp edges on mirrors, prisms, or glass plates. Fire polish the ends of glass tubing.
- ☐ When inserting tubing through stoppers, lubricate the glass with petroleum jelly. Push the tubing away from you, never toward you.
- ☐ Keep a broom and dustpan at hand for sweeping up breakage.
- ☐ Never drink or eat from glassware used for an experiment.
- ☐ Don't heat the bottom of glass test tubes. Turn them sideways, and point them away from people—including yourself!

Fire

- ☐ Keep a fire extinguisher and a fire blanket nearby whenever using flammable materials or devices.
- ☐ Have the right kind of fire extinguisher available. Electric fires can be spread by the wrong kind of extinguisher.
- ☐ Know how to call the fire department. (Phone 911 in an emergency, if 911 service is available where you live.)

Chemicals

- ☐ Wash your hands before, during, and after working with chemicals.
- ☐ Never taste, touch, or smell an unknown chemical.
- ☐ Never mix chemicals just to see what will happen.
- ☐ Add acid to water. Never add water to acid.
- ☐ Never store chemicals that might react together—such as acids and bases—in the same place.
- ☐ Never store chemicals near electrical switches or heat sources.
- ☐ Keep chemicals out of direct sunlight.

- ☐ Store hazardous chemicals in a locked, metal cabinet. Make sure the storage area is cool and dry and that nothing is leaking. Store chemicals well away from sunlight, electrical equipment, or heat sources. Don't refrigerate chemicals unless instructed to do so.
- ☐ Label all chemicals. Never trust your memory.
- ☐ Don't use mercury thermometers. Alcohol thermometers are safer. If you must use a mercury thermometer, purchase a safe clean-up kit in case the thermometer breaks and the mercury (which is highly toxic) spills.
- ☐ Know the number of your local poison crisis center before you begin.
- ☐ Never pour out more of a chemical than you need.
- ☐ Never return unused chemicals to their original jar or bottle. You'll contaminate the supply.
- ☐ Buy chemicals in small quantities, and store them in small bottles.
- ☐ Use chemicals in dry, cool, clean areas with adequate ventilation.
- ☐ Always wear a mask, safety goggles, rubber gloves, and an acid-resistant apron when working with hazardous chemicals.
- ☐ Be especially careful when moving chemicals from one place to another.

Health and hygiene

- ☐ Never share safety glasses, stethoscopes, pipettes, or any other devices that contact ears, eyes, or the mouth with another person without sterilizing the equipment thoroughly. (Ask your teacher how.)
- ☐ Wear sturdy shoes, shirts, and protective clothing in the field. Wear protective goggles, aprons, masks, and gloves in the laboratory.

SIMPLE SUBSTITUTES

If you don't have THIS...	Try THIS...
Filter paper	Paper towel or coffee filter
Autoclave	Pressure cooker or oven
Balance	Diet or kitchen scale
Stopwatch	Kitchen timer or clock with second hand
Forceps	Barbecue or kitchen tongs; tweezers
Stirring rods	Chopsticks
Plastic or rubber tubing	Aquarium tubing; flexible straws
Flasks/beakers	Jars from mayonnaise, jam, baby food, and so on
Test tube	Olive jar
Graduated cylinder	Measuring cup/spoons
Lab thermometer	Aquarium temperature strip or fever thermometer (if temperature ranges will work)
Dropper bottle	Plastic lemon or lime (emptied, washed, and rinsed)
Laboratory weights	Fishing weights
Gears and belts	Cribbage board or Peg Board, spools, and rubber bands

SUPPLIES AND MATERIALS YOU THOUGHT YOU COULDN'T GET

Don't assume you can't get the materials or equipment you need for your project. The following are but a few of the treasures available from scientific supply houses, craft and hobby centers, sports shops, pet shops, drugstores, garden supply shops, and hardware stores. You can often borrow equipment and supplies from colleges, universities, or industry. Perhaps learning to use a new apparatus or work with a new organism will give you a great idea for an original science project.

- ☐ Glucose test strip
- ☐ Fever thermometer
- ☐ Laboratory thermometer
- ☐ Temperature strip thermometer
- ☐ Stopwatch
- ☐ Cultures of bacteria
- ☐ Medium (food) for growing bacteria
- ☐ Protozoan cultures
- ☐ Microscope
- ☐ Model rocket kits
- ☐ Moths, butterflies, and ants
- ☐ Incubator for eggs
- ☐ Light meter
- ☐ Sound meter
- ☐ Lenses
- ☐ Polarizing light filters
- ☐ Cloud chamber
- ☐ Compressed gases
- ☐ Electrical supplies including switches, outlets, galvanometers, ammeters, and voltmeters
- ☐ Centrifuge
- ☐ Pulleys
- ☐ Magnets
- ☐ Compass
- ☐ Unusual plants and seeds
- ☐ Antibiotic disks

- ☐ Plankton net
- ☐ Safety equipment including goggles and gloves, earplugs, face mask, fire extinguishers, smoke detectors, and rubber aprons
- ☐ Electronics equipment
- ☐ Telescope
- ☐ Psychrometer and/or hygrometer (to measure humidity)
- ☐ Anemometer (wind speed indicator)
- ☐ Barometer (to measure atmospheric pressure)
- ☐ Microscope slides
- ☐ Forensics kits for fingerprints, document analysis, hair analysis, and fiber identification
- ☐ Lasers
- ☐ Audiometer
- ☐ Kits for testing soil
- ☐ Kits to determe biodiversity (the number of different kinds of plants/animals in an area)
- ☐ Test kits to detect air pollutants including chemicals such as carbon monoxide, carbon dioxide, and hydrocarbons
- ☐ Kits to test water including alkalinity, turbidity, hardness, dissolved solids, phosphorous, nitrate, and many more
- ☐ Fern spores for all kinds of projects ranging from life cycle studies to inheritance projects
- ☐ Kits for growing crystals

THE METRIC SYSTEM

Generally, it's best to do what scientists do: make all your measurements in metric. The basic unit of measure for

- length is the meter (m)
- volume is the liter (l)
- weight is the gram (g)
- temperature is the degree Celsius (°C)

If you use metric tools for your measurements, you'll have no need to convert. If, however, you find yourself stuck with measurements in the English system, you can convert by using some simple arithmetic:

To change...	To...	Multiply by:
Inches	Centimeters	2.54
Inches	Millimeters	25.4
Feet	Centimeters	30.5
Feet	Millimeters	305
Yards	Meters	0.914
Square yards	Square meters	0.836
Miles	Kilometers	1.61
Ounces (solid, not fluid)	Grams	28.3
Pounds	Kilograms	0.45
Quarts*	Liters	0.95
Cups*	Milliliters	250
Teaspoons*	Milliliters	5
Cubic inches	Cubic centimeters	16.387

To change...	To...	Multiply by:
Cubic inches*	Milliliters	16,387
Acres	Square meters	4,047
Square miles	Square kilometers	2.59
Square miles	Hectares	259

*This conversion is valid for liquids only. Solids, such as sugar or salt, must be weighed on a scale in grams or kilograms (or weighed in ounces/pounds and converted to grams/kilograms).

Can't find it here? Check the dictionary under "Metric System" (in the *M*s) and "Weights and Measures" (in the *W*s).

LOOKING FOR HELP? HAVE YOU CONSIDERED THESE EXPERTS?

- ☐ Therapists: speech, occupational, physical, or recreational
- ☐ Dog and horse breeders/trainers
- ☐ Exterminators
- ☐ Farmers
- ☐ Sports center/health club staff
- ☐ Construction workers
- ☐ Mechanics
- ☐ Engineers
- ☐ Cooks
- ☐ Dentists
- ☐ Pharmacists
- ☐ University or college teachers
- ☐ Hobby shop clerks
- ☐ Hobby club members
- ☐ Telephone/cable workers
- ☐ Clinic/hospital personnel
- ☐ Health department workers
- ☐ Zookeepers and caretakers
- ☐ Aquarium supply shop workers
- ☐ Pet shop workers
- ☐ Garden center workers
- ☐ Sports enthusiasts
- ☐ Museum curators
- ☐ Computer clubs
- ☐ Bird watchers
- ☐ Scouting organizations
- ☐ Community service clubs/volunteers
- ☐ Police officers
- ☐ Local horticultural societies, garden clubs, or botanical gardens
- ☐ Local chapters of conservation clubs

CHECKLIST FOR TABLES AND GRAPHS

Before completing your report and display, make sure each table and graph has

- ☐ A title
- ☐ A number with each title: Number tables in series, such as Table 1, Table 2, and so on; number graphs in sequence, such as Figure 1, Figure 2, and so on
- ☐ The correct unit of measurement specified for each variable (for example grams, millimeters, °C, and so on)
- ☐ Results from several trials, if appropriate
- ☐ Meaningful measurements and useful calculations
- ☐ Been checked and double-checked for accurate calculations

Make sure graphs have

- ☐ The dependent variable on the vertical axis (if not a circle graph)
- ☐ The independent variable on the horizontal axis (if not a circle graph)
- ☐ Variables labeled, with units, on both the horizontal and vertical axes
- ☐ Correct, equally divided scales on both axes
- ☐ Scales beginning at zero
- ☐ Correctly plotted data points
- ☐ An appropriate graphing technique—line, bar, or circle
- ☐ A legend, if needed

DO NOT DISPLAY

Many science fairs do not allow you to display

- ☐ Living organisms
- ☐ Dried plant materials
- ☐ Taxidermy specimens or parts
- ☐ Preserved vertebrate or invertebrate animals, including embryos
- ☐ Human or animal food
- ☐ Human or animal parts except hair, teeth, nails, dried animal bones, histological dry-mount sections, and wet-mount tissue slides
- ☐ Soil or waste samples
- ☐ Chemicals, including water, poisons, drugs, controlled substances, or hazardous substances or devices
- ☐ Dry ice or other sublimating solids (solids that change directly into a gas)
- ☐ Sharp objects
- ☐ Flames or highly flammable materials
- ☐ Empty tanks that previously contained combustible liquids or gases, unless the tanks have been purged with carbon dioxide
- ☐ Batteries with open-top cells
- ☐ Firearms, ammunition, or explosives
- ☐ Awards, medals, business cards, flags, or other symbolic objects
- ☐ Photographs or drawings of vertebrate animals living in other-than-normal conditions

In many science fairs, you may display but not operate

- ☐ Unshielded belts, pulleys, chains, or moving parts with tension or pinch points
- ☐ Class III or IV lasers
- ☐ Any device requiring more than 110 volts of electricity

In many science fairs, you may display and operate, with certain restrictions (check with your teacher or science fair organizer)

- ☐ Electrical appliances and electronic devices, in general, but wires, switches, and exposed metal parts connected to an electrical source must be out of reach of viewers and covered to prevent injury; cover all moving parts and wire connections
- ☐ Class II lasers
- ☐ Shielded vacuum tubes or ray-generating devices
- ☐ Secured, pressurized tanks containing noncombustible materials
- ☐ Insulated apparatuses producing extremes of heat or cold
- ☐ Properly wired and insulated 110-volt AC circuits

Check county, state, and federal laws and regulations regarding wiring, toxicity, fire hazards, and general safety.

Advance approval is required for any project that involves

- ☐ Animals
- ☐ Humans
- ☐ Pathogenic (disease-causing) agents or organisms
- ☐ Recombinant DNA

HOW CAN I HELP? A CHECKLIST FOR PARENTS, TEACHERS, AND MENTORS

Do

- ☐ Help with budget and finances, if possible and if needed
- ☐ Provide transportation to libraries and meetings with experts
- ☐ Teach techniques such as how to do library research, write letters, or access the Internet
- ☐ Help with the reading of difficult research papers
- ☐ Assist in locating materials, supplies, and equipment
- ☐ Give advice *sparingly*
- ☐ Offer constructive criticism *when requested*
- ☐ Suggest solutions to problems *when invited*
- ☐ Insist on full and adequate safety precautions
- ☐ Listen
- ☐ Encourage
- ☐ Ask probing questions (instead of giving answers)
- ☐ Look for the best in whatever has been done
- ☐ Be a role model of the joy in asking questions and seeking answers

Don't

- ☐ Design the project
- ☐ Collect the data
- ☐ Write the report
- ☐ Build the display
- ☐ "Bail out" the student at the last minute
- ☐ Do any "favors" that take control of the project away from the student!

ORGANIZATIONS AND AGENCIES

Whatever your interest, an organization, society, agency, or club probably shares it. Listed here are only a few of the thousands of government offices, professional societies, and special interest groups that may help you with your project. Many provide pamphlets, brochures, and fact sheets. Some maintain lists of experts willing to help. Some will accept e-mail inquiries.

If you don't find what you need here, ask your librarian. In the reference section of most libraries, you'll find directories of organizations, foundations, and corporations.

Professional societies

American Association for the Advancement of Science
1200 New York Avenue NW
Washington, DC 20005
http://www.aaas.org

American Association of Physics Teachers
One Physics Ellipse
College Park, MD 20740
http://www.aapt.org

American Chemical Society
1155 16th Street NW
Washington, DC 20036
http://www.acs.org

American Geophysical Union
2000 Florida Avenue NW
Washington, DC 20009
http://earth.agu.org

American Institute of Biological Sciences
1444 Eye Street NW
Suite 200
Washington, DC 20005
http://www.aibs.org

American Physical Society
One Physics Ellipse
College Park, MD 20740
http://www.aps.org

American Psychological Association
750 First Street NE
Washington, DC 20002
http://www.apa.org

American Society of Agronomy
677 South Segoe Road
Madison, WI 53711
http://www.agronomy.org

American Society for Microbiology
1325 Massachusetts Avenue NW
Washington, DC 20005
http://www.asmusa.org

American Society of Plant Physiologists
15501 Monona Drive
Rockville, MD 20855
http://www.aspp.org

American Veterinary Medical Association
1931 North Meacham Road, Suite 100
Schaumburg, IL 60173
http://www.avma.org

Geological Society of America
3300 Penrose Place
Boulder, CO 80301
http://www.geosociety.org

Institute of Electrical and Electronics Engineers
P.O. Box 1331
445 Hoes Lane
Piscataway, NJ 08855
http://www.iee.org

National Association of Biology Teachers
11250 Roger Bacon Drive #19
Reston, VA 20190
http://www.nabt.org/index.html

National Science Teachers Association
1840 Wilson Boulevard
Arlington, VA 22201
http://www.nsta.org

National Society of Professional Engineers
1420 King Street
Alexandria, VA 22314
http://www.nspe.org

The Society for Integrative and Comparative Biology
401 N. Michigan Avenue
Chicago, IL 60611
http://www.sicb.org

U.S. government agencies

Check the phone book for your town and for your state capital. You'll find numerous local and state agencies similar to the federal ones. City, county, and state health departments are good sources of information, as are fish and wildlife services and departments of agriculture.

Centers for Disease Control
1600 Clifton Road NE
Atlanta, GA 30333
http://www.cdc.gov

Federal Maritime Commission
800 North Capitol Street NW
Washington, DC 20573
http://www.fmc.gov

National Aeronautics and Space Administration
Education Division, NASA Headquarters
Washington, DC 20546
http://www.nasa.gov

National Institutes of Health
9000 Rockville Pike
Bethesda, MD 20892
http://www.nih.gov

National Oceanic and Atmospheric Administration
U.S. Department of Commerce
NOAA Public and Constituent Affairs
Room 6013
14th Street & Constitution Avenue NW
Washington, DC 20230
http://www.noaa.gov

Smithsonian Institution
Smithsonian Information
SI Building, Room 153, MRC 010
Washington, DC 20560
http://www.si.edu

U.S. Department of Agriculture
14th & Independence Avenue SW
Washington, DC 20250
http://www.usda.gov

U.S. Department of Justice
Environment and Natural Resources Division
950 Pennsylvania Avenue NW
Washington, DC 20530
http://www.usdoj.gov

U.S. Environmental Protection Agency
Public Information Center
401 M Street SW
Washington, DC 20460
http://www.epa.gov

U.S. Fish and Wildlife Service
4401 North Fairfax Drive
Webb 300
Arlington, VA 22203
http://www.fws.gov

Special interest groups

For every concern of health, environment, or public policy, a special interest group probably exists that seeks to promote its members' point of view. You can find those active in your area by checking the newspaper, asking people you know, or looking in the phone book. Also, you may contact the national headquarters of an organization to find your local chapter.

The following list includes only a few of the many hundreds of organizations supporting environmental protection and conservation.

Environmental Defense Fund
257 Park Avenue South
New York, NY 10010
http://www.edf.org

Greenpeace
1436 U Street NW
Washington, DC 20009
http://www.greenpeaceusa.org

National Audubon Society
700 Broadway
New York, NY 10003
http://www.audubon.org

National Wildlife Federation
8925 Leesburg Pike
Vienna, VA 22184
http://www.nwf.org

Sierra Club
85 Second Street, Second Floor
San Francisco, CA 94105
http://www.sierraclub.org

World Wildlife Fund
1250 Twenty-Fourth Street NW
Washington, DC 20037
http://www.worldwildlife.org

SCIENTIFIC SUPPLY HOUSES

Science supply houses provide the materials, supplies, and equipment you need for your project. Their catalogs can also be good sources of ideas and their customer service representatives excellent sources of help. Most science teachers keep a supply of catalogs handy. Check those catalogs first. Then contact one of these businesses.

AccuLab Products Group
614 Scenic Drive, Suite #104
Modesto, CA 95350
http://www.sensornet.com

Arbor Scientific
P.O. Box 2750
Ann Arbor, MI 48106
http://www.arborsci.com

Blue Spruce Biological Supply
701 Park Street
Castle Rock, CO 80104
http://www.bluebio.com

Carolina Biological Supply Company
2700 York Road
Burlington, NC 27215
http://www.carolina.com

Central Scientific Co.
3300 Cenco Parkway
Franklin Park, IL 60131
http://www.cenconet.com

Connecticut Valley Biological Supply Co., Inc.
82 Valley Road
P.O. Box 326
Southampton, MA 01703

Cuisenaire Co. of America
10 Bank Street
P.O. Box 5026
White Plains, NY 10602-5026
http://www.cuisenaire.com

Delta Education, Inc.
12 Simon Street
P.O. Box 3000
Nashua, NH 03061
http://www.delta-ed.com

Edmund Scientific Co.
101 E. Gloucester Pike
Barrington, NJ 08007
http://www.edsci.com

Estes Industries
1295 H Street
Penrose, CO 81240
http://www.service.com/estes/estes.html

Fisher Science Education
Educational Materials Division
485 South Frontage Road
Burr Ridge, IL 60521
http://www.fisheredu.com

Frey Scientific
Division of Beckley Cardy
100 Paragon Way
Manfield, OH 44903
http://www.BeckleyCardy.com

Lamotte Co.
P.O. Box 329
Chestertown, MD 21620
http://www.lamotte.com

Nasco Science
901 Janesville Avenue
P.O. Box 901
Fort Atkinson, WI 53538

Northwest Laboratory Supply
5570 Neilson Road, Unit B
Ferndale, WA 98248

Northwest Scientific Supply
171 Cooper Avenue
Tonawanda, NY 14150
http://visual.net/home/nwscience

PASCO Scientific
10101 Foothills Boulevard
Box 619011
Roseville, CA 95678-9011
http://www.pasco.com

Schoolmasters Science
745 State Circle
P.O. Box 1941
Ann Arbor, MI 48106
http://www.schoolmasters.com/sciat.html

Science Kit and Boreal Laboratories
777 East Park Drive
Tonawanda, NY 14150
http://sciencekit.com

VWR Scientific Products
P.O. Box 5229
Buffalo Grove, IL 60089
http://www.Sargentwelch.com

Ward's Natural Science Establishment Inc.
5100 W. Henrietta Road
P.O. Box 92912
Rochester, NY 14692-9012
http://www.wardsci.com

INDEX

NOTES